PRACTICE OF THIN LAYER CHROMATOGRAPHY

Practice of Thin Layer Chromatography

JOSEPH C. TOUCHSTONE
MURRELL F. DOBBINS
University of Pennsylvania
School of Medicine

A Wiley-Interscience Publication

JOHN WILEY & SONS
New York • Chichester • Brisbane • Toronto

Copyright © 1978 by John Wiley & Sons, Inc.

All rights reserved. Published simultaneously in Canada.

Reproduction or translation of any part of this work beyond that permitted by Sections 107 or 108 of the 1976 United States Copyright Act without the permission of the copyright owner is unlawful. Requests for permission or further information should be addressed to the Permissions Department, John Wiley & Sons, Inc.

Library of Congress Cataloging in Publication Data:

Touchstone, Joseph C
  Practice of thin layer chromatography.

  "A Wiley-Interscience publication."
  Includes index.
    1. Thin layer chromatography.   I. Dobbins, Murrell F., joint author.   II. Title.

QD79.C8T68   544'.92   77-22075
ISBN 0-471-88042-6

Printed in the United States of America

10 9 8 7 6 5 4 3 2

This book is dedicated to our wives,

Phyllis and Clare

Preface

There are several widely used texts on thin layer chromatography
(TLC) each of which is composed of two major sections: one section dealing with basics; the other section dealing with specific
compound class applications. In most instances, the chapters
dealing with the basics of TLC do not cover their topics in enough
depth and detail to enable the reader to carry out the operation
described with confidence and success.

In this volume, we have attempted to describe in detail all
the basic operations necessary for successful TLC in such a way
that the reader may readily carry them out without hesitation.
Rather than presenting complete literature reviews in each of the
topic areas, we have drawn information from practical experience
and from the literature describing practical methods. For best
overall results, the beginning TLC practitioner should read this
book from beginning to end before beginning any chromatography in
order to familiarize himself or herself with the entire TLC process. This familiarity will allow each step to proceed more
smoothly than might otherwise be possible.

Using this information, the worker in TLC can readily perform
the desired separations in his or her own laboratory.

This volume would not have been possible without the generous
support of the Department of Obstetrics and Gynecology, School of
Medicine, the University of Pennsylvania. Our thanks are also extended to the firms and individuals who granted permission to use
graphs and illustrations of equipment and procedures. The inclu-
sion of equipment from a given supplier should not be construed as
a recommendation, nor should failure to cite any product or sup-

plier be construed as a lack of recommendation.

We also appreciate the devoted work of those who diligently worked to assemble the data: Mr. Philip Blackwood, Mrs. Clare Dobbins, and Miss Cheryl Crowder. A very special thanks to the late Mr. Eric Nyberg for constructive criticism of the manuscript.

Joseph C. Touchstone
Murrell F. Dobbins

Philadelphia, Pennsylvania
June, 1977

Contents

xi

CHAPTER 14   THE COMBINATION OF THIN LAYER CHROMATOGRAPHY WITH
OTHER ANALYTICAL TECHNIQUES

xiv

Glossary of TLC Terms

absorbent	Substance that absorbs others, usually on the basis of wetting ability.
activation	The process of heating a TLC plate to drive off moisture resulting from layer preparation or adsorption from the atmosphere.
adsorbent	Substance adhering to another due to attraction between surface atoms of the two substances resulting from intermolecular forces such as hydrogen bonding, electrostatic forces, and charge-transfer forces.
adsorption chromatography	Process whereby a sample is separated by interaction between adsorptive forces of a medium (stationary phase) and a solvent (mobile phase).
alumina	Common adsorbent; Al_2O_3
argentation TLC	TLC employing silver nitrate impregnated in the layer material, usually silica gel. This impregnation changes the separation characteristics of the silica gel.

ascending chromatography	Chromatography in which the mobile phase moves upwards in the medium.
bed	A column or layer of porous material containing the stationary phase, the interstices being filled with mobile phase.
band	Chromatographic zone; region where the separated substance is concentrated.
CC	Column chromatography.
cellulose	Common medium for separation on a TLC plate.
chamber	Tank, jar, or vessel in which chromatographic separation takes place.
chamber saturation	Equilibration of the chamber or tank with mobile phase before the plate is placed into it.
chromatogram	A series of separated bands or zones in or on the stationary phase. The end product of the chromatography process.
chromatographic solvent	Solvent or mixture of solvents used as the mobile phase.
chromatographic system	Combination of the solvent, the sorbent, and components of the sample mixture. The interactions of the system determine the selectivity of the separation.
chromatography	A method of analysis in which the flow of a mobile phase (gas or liquid) promotes the separation of substances by differential migration from a narrow initial zone in a sorptive medium.
chromatoplate	A thin layer plate; a layer of sorbent coated on a solid support such as glass, aluminum, or plastic.

continuous development	Development occurring over a distance that is usually greater than one plate length. Development is often expressed as a function of time rather than distance.
deactivation	The process of making the chromatographic layer less active to decrease its separation capabilities. Usually done with water.
densitometry	Measurement of a zone on a layer with an instrument that determines the density of the zone.
descending chromatography	Chromatography in which the mobile phase moves downwards in the sorptive medium.
destructive detection.	A detection process that changes the chemical nature of the substance being detected in an irreversible manner. Sulfuric acid charring is one example.
detection	The process of locating a separated substance on a chromatogram, whether by physical methods, chemical methods, or biological methods.
developing solvent	Mobile phase.
development	The movement of mobile phase in the chromatogram to effect separation.
diatomaceous earth	A naturally occurring fine white powder, formed from the skeletons of microscopic marine organisms. Also called kieselguhr.
distribution coefficient	$$k = \frac{\text{amount of solute per unit of stationary phase}}{\text{amount of solute per unit of mobile phase}}$$
eluent	Solvent that removes a sample from a medium.

eluotropic series

Series of solvents or solvent mixtures arranged in order of eluting power.

elution

Removal of a solute from a sorbent by passage of a suitable solvent.

flat-bed chromatography

Common term for thin layer or paper chromatography occurring in a single plane. Sometimes called planar chromatography.

front

The visible boundary at the junction of the mobile-phase wetted layer and the "dry" layer. If a trough chamber is used with equilibration for development, the "dry" layer can contain amounts of the mobile phase.

GC (GLC)

Gas (liquid) chromatography.

gradient TLC

Separation on a sorbent layer that has changing characteristics, that is, a gradient, from one portion of the layer through an adjoining portion of the layer.

gradient elution

Development using a solvent system whose composition is continuously changing to effect separation; normally done to increase the strength of the eluent.

HPLC

High pressure liquid chromatography.

HPTLC

High performance thin layer chromatography.

hR_f

$100 \times R_f$

impregnation

Loading of the sorbent with a liquid or a solid to change the chromatographic behavior of the layer. An example is $NaNO_3$ impregnated silica gel.

in situ

Occurring in place, e.g., on the thin layer.

ion exchange	Process whereby ions of the same charge replace each other in a given phase. In chromatography, it usually refers to symptoms where the stationary phase is made of an ionic polymer, which can be a synthetic resin or a specially treated mineral.
IR	Infrared spectrometry.
kieselguhr	Diatomaceous earth.
migration	Travel of sample in the medium in the direction of the mobile phase.
mobile phase	The moving phase (solvent or gas) of a chromatographic system.
MS	Mass spectrometry.
multiple chromatography	Chromatography repeated a number of times using the same or different mobile phases.
nondestructive detection	Detection of a substance on a chromatogram by a process that will not permanently change the chemical nature of the substance being detected. Visualization with iodine vapor is one example of a nondestructive method.
origin	Point where sample is applied.
partition	Divide or distribute between.
partition chromatography	Process in which sample is separated by partition between two liquid phases or between a gas and a liquid. One liquid is stationary while the other is mobile.
partition coefficient or ratio (K_d)	Ratio of concentration of solute after partition between two immiscible phases.

$K_d = \frac{C_s}{C_m}$, where C_s and C_m are the concentrations in the stationary and mobile phases, respectively.

PC   Paper chromatography.

PLC   Preparative layer chromatography. Used for the separation of larger amounts of substance than are normally separated with regular, analytical TLC. Normally a thicker layer (500-2000 μ) or sorbent is employed than in TLC.

PMD   Programmed multiple development. The repeated development of a TLC plate with the same mobile phase in the same direction for gradually increasing distances.

polar   Highly charged, or with uneven electricalycharges. Degrees of solubility in water can be used as a measure of polarity. In organic chemistry a polar molecule is one with a large dipole moment. In chromatography a polar molecule is one whose distribution coefficient favors the polar phase. Affinity of substances for polar solvents depends on their dipole moments and their molecular volumes. It is clear that polarity in the strict sense is not always synonymous with solubility.

precoated plates or sheets   Commercially available than layer plates or sheets ready for use in TLC.

resolution   The degree of separation between two substances expressed as

$$Rs = \tfrac{1}{4}(\alpha - 1) \sqrt{N} \left(\frac{k'}{k'+1}\right),$$

where α is the separation factor or

the ratio of the capacity factors between two solutes k_1/k_2; N is the number of theoretical plates in the sorbent bed; k' is the average of k_1 and k_2. The capacity factor k is the equilibrium ratio of total solute in the stationary phase to total solute in the mobile phase.

reversed-phase chromatography — Chromatography on a sorbent impregnated with a nonpolar and nonvolatile liquid as a stationary phase. Separation is effected by a polar mobile phase.

R_f value — A ratio: the distance from the origin to the center of the separated zone divided by the distance from the origin to the solvent front.

R_m value — $\text{Log } \left(\dfrac{1}{R_f} - 1 \right)$

sandwich chamber — Developing chamber formed from the plate itself, a spacer, and another nonlayered cover plate that stands in a trough containing the mobile phase.

secondary front — An additional solvent (mobile phase) front, lower than the primary solvent front. Occurs because components of the mobile phase have demixed and migrated apart from the other components.

silica gel — Silicic acid. The most widely used sorbent for TLC. Also used in column chromatography.

solvent — Liquid used for mobile phase. Not identified *a priori* with mobile phase.

solvent front — The forwardmost point of the mobile phase during development.

sorbent	A generalized term for the chromatographic stationary phase in which the nature of the force (adsorption, ion exchange, or reversed phase) is not specified.
starting point line	Position on chromatogram where the sample is applied. Usually 10-20 mm from the bottom of the plate. Also called the origin.
stationary phase	The phase of the chromatographic system that is made up of the surface of an adsorbent or liquid held by the support of a partition or gel system.
stepwise elution	Development using an eluent whose composition is changed using discontinuous, stepped gradients, in contrast to *gradient elution*.
support	The sheet of glass, plastic, or aluminum upon which the TLC sorbent is coated. Gives physical strength to the layer.
tailing	Incomplete separation of zones, often resulting in elongation of a zone.
TLC	Thin layer chromatography (chromatogram).
TLG	Thin layer gel chromatography employing a gel, such as Sephadex, coated on a glass plate for the separation of molecules predominantly according to their size.
two-dimensional chromatography	Successive development of a chromatogram in directions orthogonal to each other with the same or different mobile phase.
two-dimensional development	A two-step development technique in which a plate is first developed in one mobile phase, then dried, turned

through 90°, and developed in a second, different mobile phase.

UV   Ultraviolet light.

zone (also spot or band)   The distribution of the solute or separated compound on the stationary phase before, during, and after chromatography.

PRACTICE OF THIN LAYER
CHROMATOGRAPHY

CHAPTER 1

Basics of Thin Layer Chromatography

1.1 INTRODUCTION

Thin layer chromatography (TLC) is one of the most popular and widely used separation techniques. The reasons for this are many, and include ease of use, wide application to a great number of different samples, high sensitivity, speed of separation, and relatively low cost. A variety of apparatus for TLC is commercially available.

TLC is used for the separation of substances in a wide variety of fields. Amino acids from food protein, hallucinogenic alkaloids from plants, steroids from the urine of a newborn infant, morphine in the blood of an overdose victim, and pesticides from soil may be separated by TLC, with sensitivities of one microgram or less. Many separations can be accomplished in less than an hour, at very reasonable cost. An initial investment of a couple of hundred dollars will purchase most of the equipment necessary to do TLC on a qualitative basis. Instrumentation for quantitative analysis on the TLC plate will cost additional thousands of dollars.

Chromatography is a method of separating a mixture into its various components. It makes use of heterogeneous equilibrium established during the flow of a solvent called the mobile phase through a fixed (stationary) phase to separate two or more components from material carried by the solvent. The stationary phase can be either solid or liquid. The mobile phase can be either liquid or gas. Thus, chromatography can be classified as (1) solid-liquid, (2) liquid-liquid, (3) gas-solid, or (4) gas-liquid.

At this point it is desirable to introduce a number of im-

1

portant concepts. The first of these is polarity. A polar compound (one with high chromatographic polarity) is one that is
held by the stationary phase, whereas a nonpolar substance tends
to move forward in the mobile phase and be held less by the
stationary phase. This term is also used to describe the solvents used as the mobile phase. Methanol, for example, is a
polar solvent, since it is extremely effective as a mobile phase
in moving substances through the stationary phase. Cyclohexane
is an example of a nonpolar solvent; it will not move many substances. This type of polarity should not be confused with the
concept used in organic chemistry, in which polarity is expressed
in terms of dipole moments. In a chromatographic sense, benzene
is a more polar solvent than cyclohexane, yet neither has a
dipole moment.

The most common form of chromatography is adsorption chromatography. Here the stationary phase is a solid such as alumina or silica gel. The sample substance to be separated is
applied at one end of this stationary phase, and the mobile
phase is allowed to flow through. In TLC, capillary action in
the finely divided stationary phase causes the mobile phase to
move. Separation occurs when one substance in a mixture is more
strongly adsorbed by the stationary phase than the other components in the mixture. Since adsorption is essentially a surface phenomenon, the degree of separation is dependent on the
surface area of the adsorbent available; hence the emphasis on
small particle size of the adsorbent.

The main factor in any form of chromatography is the distribution coefficient of a substance between the two phases of
the system in use, where

$$
\text{Distribution coefficient (k)} = \frac{\text{amount of solute per unit of stationary phase}}{\text{amount of solute per unit of mobile phase}}
$$

Usually in adsorption chromatography, k is dependent on the temperature and the concentration of the solute. At a given temperature, the relationship between the amount of solute in each
phase can be expressed graphically as the adsorption distribution
isotherm. The ideal isotherm is a 45° linear curve obtained
from plotting the concentration of the solute in the stationary
phase versus its concentration in the mobile phase.

A nonlinear isotherm in a given TLC system is exemplified by
the shape of the separated zone (spot) on the plate. Spots have
either a well-defined front half with a tail, appearing as an
upside-down teardrop, or a well-defined back half, appearing as a
normal teardrop, depending on whether the distribution isotherm
is convex or concave, respectively.

It is usually possible with TLC to work at such concentrations that the distribution isotherms approach linearity. If two substances in a mixture have different isotherms, they can be separated. The movement of each substance in a given system is a characteristic of the particular substance and can be used for qualitative identification.

Partition chromatography is also used in the separation of many compounds. An equilibrium between two phases is involved; one of the phases is fixed, that is, it is held by an inert support such as cellulose or kieselguhr; the other phase is mobile. In TLC, layers usually contain water adsorbed onto the surface during manufacturing or exposure to the atmosphere. This water will behave as the stationary liquid phase, partitioning the sample components between it and the mobile phase. The support material in such a case is not directly involved with the separation. The equilibrium occurs between the stationary liquid and the liquid mobile phase. The separation of a compound takes place on the basis of nonbonding interactions in the phases. The chief advantage of partition over adsorption chromatography is that the distribution isotherms are reasonably linear over a fairly wide concentration range.

Thin layer chromatography can be used: (1) to simply check the purity of a substance, (2) to attempt to separate and identify the components in a mixture, or (3) to obtain a quantitative analysis of one or more of the components present.

From a purely experimental standpoint, chromatography can be carried out in columns or layers. In column chromatography the adsorbent or medium is packed into a tube. With modern columns it is possible to reproduce separations because of uniformity in particle size and characteristics. The sample is introduced into the top of the column, and the liquid or gas is allowed to flow through. In the case of thin layer separations, the sample is applied at one end of the layer, and this end is immersed in the mobile phase with the sample just above the liquid level. The tank used for this operation is usually rectangular. After the front of the mobile phase has reached to within a short distance from the top of the layer, the plate is removed and dried prior to detection procedures.

1.2 HISTORY

Thin layer chromatography was first referred to in 1938 by two Russian workers, Izmailov and Shraiber (1) in what they called drop chromatography on horizontal thin layers. Little notice was made of the method until 10 years later, when two American chem-

ists described the separation of terpenes in essential oils by
thin layer chromatography (2). Thin layer chromatography, as it
is presently known, began to attract attention through the work
of Kirchner and his associates, starting in 1951 (3-5). The procedure was not generally accepted in its early years because
available media and apparatus for coating the plates lacked uniformity.

It was not until 1958, when Stahl (6) described equipment
and efficient sorbents for the preparation of plates, that the
effectiveness of the technique for separation was shown.

The method is now one of the most frequently described separation techniques in qualitative as well as quantitative analysis. Stahl's book, *Thin Layer Chromatography* (7), first appeared in 1965. This work and Kirchner's *Thin Layer Chromatography* (8), among others, gave the real impetus to development in
the field. In its second edition, published in 1969, Stahl's
book (9) was greatly expanded.

1.3 TLC PROCEDURE

Thin layer chromatography is a separation method in which uniform
thin layers of sorbent or selected media are used as a carrier
medium. The sorbent is applied to a backing as a coating to obtain a stable layer of suitable size. The most common support is
a glass plate, but other supports such as plastic sheets and
aluminum foil are also used. The four sorbents most commonly
used are: silica gel, alumina, kieselguhr (diatomaceous earth),
and cellulose.

Silica gel (silicic acid), is the most popular layer material. It is slightly acidic in nature. In order to hold the silica gel firmly on the support, a binding agent such as plaster of
Paris (calcium sulfate hemihydrate) is commonly used. This binding agent may be omitted if the silica gel employed has a very
small particle size. Fine particles will adhere well to the support without a binder.

Two ultraviolet (UV) indicators, to aid in the location of
separated substances, are also incorporated, either singly or
together, in silica gel or other layer materials. Sodium fluorescein fluoresces when exposed to ultraviolet light of 254 nm
(mμ) wavelength, so that substances absorbing this wavelength
will contrast sharply by appearing dark while quenching the
greenish-yellow fluorescing background. The sodium salts of
hydroxy-purene-sulfonic acids fluoresce at 366 nm and provide a
contrasting background for substances that absorb at this fre-
quency.

Alumina (aluminum oxide) is also widely used as a sorbent. Chemically it is basic, and for a given layer thickness it will not separate quantities of material as large as can be separated on silica gel. Alumina is more chemically reactive than silica gel, and care must be exercised with some compounds and compound classes to avoid decomposition or rearrangement of these substances during sample application, storage before development, or development.

Diatomaceous earth (kieselguhr) is a chemically neutral sorbent that does not separate or resolve as well as either alumina or silica gel. It is used mainly as the support for the stationary phase in partition chromatography. It is not generally available with fluorescent indicator.

Cellulose is used as a sorbent in TLC when it is convenient to perform a given paper chromatographic separation by TLC in order to decrease the amount of time necessary for the separation and increase the sensitivity of detection. Many separations achieved by paper chromatography (PC) can be directly transferred to TLC on cellulose. This is important when only a small amount of sample is available, as TLC does not generally require as much sample as PC. The primary separation mechanism is partition, where the cellulose becomes a support for a stationary phase of water adsorbed from the atmosphere. The mobile phase is that put into the tank and used for development. Because of its fibrous, tenacious nature, cellulose is usually coated onto a plate without a binder. It is available with and without fluorescent indicator.

Other substances used as sorbents include a variety of ion-exchange cellulose powders, polyamide powder, and Florisil. Commercially available coated plates employing these sorbents are listed in Chapter 3. Modern sorbent materials are of uniform particle size and characteristics, and it is possible to reproduce the quality of a layer.

The term sorbent is used in a general sense to include layer materials that may be used for either adsorption or partition chromatography. Table 1.1 lists most of the common sorbents, the major form of separation that occurs when they are used, and typical compound applications for each. Some generalizations may be made about their use. Adsorption is used to separate nonpolar, hydrophobic (non-water-soluble) substances with nonpolar mobile-phase solvents. Partition may be used for polar, hydrophilic (water-soluble) substances, with polar mobile-phase solvents. See Table 1.2.

TABLE 1.1   SORBENT MATERIALS AND MODE OF SEPARATION

Sorbent	Chromatographic Mechanism	Typical Applications
Silica gel	adsorption	steroids, amino acids, alcohols, hydrocarbons, lipids, aflatoxins, bile acids, vitamins, alkaloids
Silica gel RP	reversed phase	fatty acids, vitamins, steroids, hormones, carotenoids
Cellulose, kieselguhr	partition	carbohydrates, sugar alcohols, amino acids, carboxylic acids, fatty acids
Aluminum oxide	adsorption	amines, alcohols, steroids, lipids, aflatoxins, bile acids, vitamins, alkaloids
PEI cellulose	ion exchange	nucleic acids, nucleotides, nucleosides, purines, pyrimidines
Magnesium silicate	adsorption	steroids, pesticides, lipids, alkaloids

The standard size for TLC plates is 20×20 cm. For most separations, the mobile phase is allowed to travel on the layer for a distance of 15 cm. Other plate sizes used are 5×20 cm, 10×20 cm, and 20×40 cm. "Micro" plates have been made from microscope slides.

In practice, a sample to be separated is applied on the layer 1-2 cm from one end of the plate. The furthermost edge of the application is called the starting point or origin. Separation is achieved by passing a solvent, the mobile phase, through the layer. The layer, with the sample zones at the bottom, is

TABLE 1.2 COMPARISON OF TLC SEPARATION MECHANISMS

	Adsorption	Partition	Reversed-Phase Partition
Compound types separated	Hydrophobic (lipophilic) of low or medium polarity	Hydrophilic inorganics and hydrophilic polar organics	Closely related hydrophobic substances
Layer type used	Activated adsorbents	Sorbent containing water, buffer, or very polar organic liquid; no activation	Sorbent containing a very nonpolar liquid stationary phase; no activation
Mobile phase	Many organics	Usually organic liquids saturated with water or buffer	Polar liquids
Common layer materials	Silica gel, alumina	Cellulose, unactivated silica gel	Cellulose, silanized silica gel
Average time for 10 cm development	20-45 min	60-90 min	60-90 min

placed on a slight angle from the vertical into a closed tank
containing a small amount of the mobile phase.

The nature and chemical composition of the mobile phase is
determined by the type of substance to be separated and the type
of sorbent to be used for the separation. The composition of a
mobile phase can be as simple as a single, pure solvent (such as
benzene used to separate dyes on alumina) or as complex as a
three- or four-component mixture containing definite proportions
of chemically different substances, such as a 1:1:1:1 solution of
n-butanol-ethyl acetate-acetic acid-water used to separate amino
acids on silica gel.

Capillary action causes the mobile phase to travel through
the medium in a process called *development*. Ascending development is the most common, but horizontal, descending, and centrifugal methods have also been reported. After the plate is dried,
the separated spots can be visualized (made visible) in a number
of ways such as viewing under an ultraviolet light or spraying
with one of a wide variety of reagents. The entire TLC process
is summarized in Figure 1.1 and will be described in detail in
succeeding chapters of this book.

The process presented in Figure 1.1 is the simplest form of
chromatography. In some respects it is the fastest. Ideally, it
also requires the least amount of equipment. Figure 1.2 shows a
TLC kit produced by one of the major TLC suppliers. Very little
is needed to perform a separation, and one does not need commercial supplies to become an experimenter in TLC. A screw-capped
jar can be substituted for the tank. Capillary pipets (e.g.,
melting point tubes) are readily obtained for applying samples in
qualitative work. You can produce your own plates by simply dipping microscope slides into a slurry. Ready-made commercial
plates are becoming less expensive, save time, and often yield
more reproducible results.

As stated in the diagram in Figure 1.1, the initial decision
to be made by the experimenter, provided the basic chemical nature of the sample is known, is which sorbent to use. The more
information that is known about the sample, the easier it will be
to choose the mobile phase and sorbent that will most likely effect the desired separation. Silica gel can be used as the sorbent for most applications. However, before the whole process
can be performed, some attention must be paid to the characteristics of the sample to be examined. The nature of the sample determines the sorbent. But often the sample must be subjected to
some previous separation such as extraction or filtration to be
suitable for satisfactory TLC. When the sample is relatively
"clean" and uncontaminated, as in the case of pharmaceutical preparations containing only two or three components, then solutions

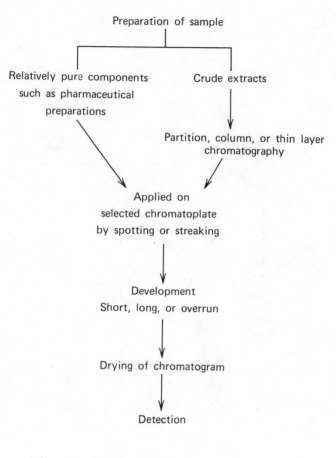

Preparation of sample

Relatively pure components
such as pharmaceutical
preparations

Crude extracts

Partition, column, or thin layer
chromatography

Applied on
selected chromatoplate
by spotting or streaking

Development
Short, long, or overrun

Drying of chromatogram

Detection

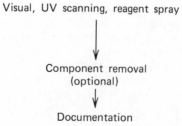

Visual, UV scanning, reagent spray

Component removal
(optional)

Documentation

Figure 1.1. The Process of Thin Layer Chromatography.

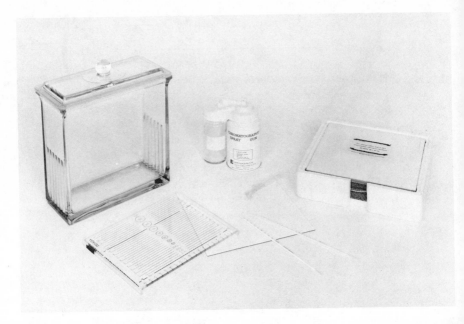

Figure 1.2. Basic equipment for thin layer chromatography. At the rear, from the left, a TLC developing tank, propellant sprayer for applying a visualization reagent to the plate, vial of spotting capillaries for applying samples, box of precoated commercial plates. In front, to the left, is a clear Plexiglas spotting template used as an aid to spotting and measuring developed spots. In the center is a thin layer plate with the spotting capillaries and measuring pipets on top. Courtesy Brinkmann Instruments.

of the sample, with proper dilution, can be applied to the TLC plate directly. However, often crude extracts of biological or industrial material are too bulky to be applied directly. These can be "prepurified" by a TLC or column separation.

The sample can be spotted or streaked on the plate. This operation can be manual or automatic using one of the many spotters or streakers now available. These are merits to both techniques. This subject will be discussed in great detail in Chapter 4. The spots must be dried before the plate is inserted in the tank.

The chromatogram can be developed for a short distance (less than half the length of the plate) or to the end of the plate; it can be developed a number of times successively with the same or

different solvents, or multidimensionally (in more than one dimension). After the chromatogram is properly dried, the spots
can be located in a number of ways. Colored zones are readily
located. Visualization under ultraviolet light can often be used.
Or the chromatogram can be treated with a wide selection of specific and nonspecific spray reagents, which will react to produce
visible colored or fluorescent substances.

After visualization, many experiments can be considered complete. If documentation is desired, the chromatogram can be
treated in a number of ways. Photography is relatively simple
and provides a permanent record. The plate can be covered with a
glass sheet and taped to seal the edges, and thus be preserved
for future reference.

The R_f value is a convenient way to express the position of
a substance on a developed chromatogram. It is calculated as the
ratio:

$$R_f = \frac{\text{distance of compound from origin}}{\text{distance of solvent front from origin}}$$

R_f values are between 0 and .999, and without units. Distance is
measured to the center of the sample zone or spot. Factors controlling R_f are discussed in Chapter 11.

If it is desired to express positions relative to the position of another substance x, the R_x can be calculated:

$$R_x = \frac{\text{distance of compound from origin}}{\text{distance of reference compound x from origin}}$$

It is possible for R_x values to be greater than 1. The R_x
is thus the "relative retention value."

According to the types of compounds being investigated, the
separations on the plate may then be evaluated by a number of instrumental techniques. If the substances are radioactive or suspected to be so, the plate can be examined by a radioisotope
scanner to locate these substances. Under proper conditions, it
is also possible to measure both the area of a spot and its density, using a photodensitometer. These characteristics are related to the quantity of a given substance in the spot and can
thereby be used to assess the amount of the substance in the sample. A quantitative calibration or reference curve should be
established beforehand using known amounts of the substance to be
quantitated.

Elution methods may also be used for quantitation and measurement of radioactivity. In this technique, the substance to
be measured in an area on the developed plate is scraped off the

plate with a spatula or blade. It is dissolved out of the layer
material (eluted) with a suitable solvent. The measurement technique is then applied to this solution.
     The purpose of this book is to describe each one of these
techniques in detail. The tools that can be used will be item-
ized, and their relative merits discussed. The sections on sorbents and solvents and methods for their selection may be the
most important. Visualization and the many detection reagents
are also covered.

1.4 OTHER FORMS OF CHROMATOGRAPHY

The use of gas chromatography, which developed almost concurrently with thin layer chromatography, is now widely appreciated.
"Modern liquid" chromatography (MLC) came into being with break-
throughs published in 1969. Thin layer chromatography is essentially the same as column chromatography and liquid chromatography, the only difference being that the medium or sorbent is in a
layer and not in a column. These procedures are distinguished by
details of equipment, materials, technique, and theoretical basis.
The term "liquid chromatography" (LC) refers to any chromatographic process in which the mobile phase is a liquid in contrast
to the moving gas phase of gas chromatography. A comparison of
the relative merits of gas chromatography and "modern" liquid
chromatography (MLC) and thin layer chromatography would illustrate some of the advantages of TLC.
     Gas chromatography (GC) has a high capability of separation
and is used for the analysis of a wide variety of mixtures. GC
can separate faster and sometimes better than other techniques.
Moreover, automated equipment is now available for trouble-free
unattended operation. GC has the advantage of readily reproducible results, since the parameters of operation such as temperature and gas flow can be controlled. The technique suffers in
that most organic compounds cannot be handled due to their insufficient volatility or lack of thermal stability.
     In contrast, TLC is not limited by the volatility or thermal
stability of the sample. Paper chromatography also enjoys this
advantage but has the disadvantage that the times involved in
achieving a separation are among the longest found in any form of
chromatography.
     The volatility or thermal stability of the sample is not a
limitation with "modern" liquid chromatography. The term "modern" LC is used here to designate the more recent developments in
instrumentation and packing materials that make high-pressure
liquid chromatography possible. Separations by liquid chromatog-

raphy can normally be done within short periods of time, although some may take hours.

One of the advantages of liquid chromatography over gas chromatography is in sample recovery. Generally, in paper chromatography, TLC, or LC, samples are readily recoverable. In column separations the mere collection of an effluent representing the desired fraction is all that is required. In paper and TLC the desired component must usually be "cut out" and eluted. The recovery of samples from GC is possible but is generally more difficult, less convenient, and not always quantitative.

In classical liquid chromatography, a column had to be packed for each separation; it could not be used repeatedly. In modern LC, columns can be used many times. In TLC the chromatogram is generally not used again. This can represent a significant expense in material and time. However, the amount of sorbent used is small, and the average cost of a TLC plate is low, whereas packings in LC are expensive. But since packings and adsorbents available for chromatography today are more uniform, it is usually possible to reproduce the required chromatogram whether it be by TLC or MLC.

Typical separations and analyses by classical LC were tedious and time consuming. MLC or TLC can be done in relatively short times, often as little as 10 min, depending on the mobile phase or the sorbent. The time factor in this controversy becomes more diverse when detection methods are considered as well as the quantitative aspects. In MLC, detection and quantitation are simple provided proper detecting instruments are available. Only recently has it been possible to monitor separations other than those detected with instruments using 254 nm and 360 nm wavelengths of the light sources. Detection in TLC can follow a number of properties depending on the compound under consideration. Absorption in the ultraviolet or visible regions can be used, or the compounds can be rendered visible by derivatization before or after the development of the chromatogram.

There has, however, been some misconception about the quantitative aspects of TLC. Early workers and even contemporaries in the field eluted the compounds from the chromatograms and then carried out the appropriate quantitative analyses. Relatively few chromatographers using TLC realized that *in situ* densitometry provides highly reproducible and sensitive methodology for quantitation of many compounds. Densitometric methodology is beyond the scope of this book; those interested should consult the recent literature on the subject (10). Many compounds can be determined at the picogram level, and determinations in the nanogram range are common.

1.5   ADVANTAGES OF TLC

TLC has a number of basic advantages over LC. While a methodology is being developed for a specific separation, TLC uses less
solvent. The polarity of the solvent or the type of solvent mixture can be changed in a matter of minutes. Little equilibration is required, and only a small amount of solvent is needed
for an exploratory chromatogram. Thus, because of short development time and easy change of the mobile phase, TLC is probably
the easiest chromatographic method to set up for a specific compound. In fact, TLC is often used to develop solvent systems for
LC.
      TLC as well as LC can be performed using adsorption, ion-ex-
change, or partition media. The choice of any of these media as
layers for TLC is limited only by the availability of the media
in a suitable form for coating the support. Many commercial suppliers have a wide selection of these chromatographic media on
precoated plates. Thus, nearly any separation is possible if the
proper combination of mobile phase and layer are used.
      Probably one of the most advantageous features of TLC as op-
posed to other chromatographic methods is the number of samples
that can be handled at one time. Traditionally, gas or liquid
chromatography are limited to the analysis of a single sample at
one time. As many as 20 samples can be applied to a single 20×20
cm TLC plate for determination at one time. This is the main
reason why TLC has been so popular in the screening of samples in
drug abuse investigations.
      The literature abounds in references to the separation of
many classes of compounds. The serious practitioner of chromatography should have ready access to the basic texts and journals. Some theory of chromatography is covered in the excellent
text of Giddings (11). The theory of LC is not much different
from that of GC and this book develops an understanding of the
fundamentals. A comprehensive review of all books on chromatography through 1969 (12) is also valuable. Books by Marini-
Bettolo (13), Bobbitt (14), Randerath (15), and Macek (16), among
others, contain much of the early literature on TLC. The bibli-
ographies in these books are quite extensive.
      Very useful is the two-volume *Handbook of Chromatography* by
Zweig and Sherma (17). This handbook lists solvent systems and
sorbents for the separation of many compounds. Also given are
the various spray reagent formulations for detection of these
compounds.
      The abstracts found in the hournal of chromatography
(Elsevier, Amsterdam) are organized according to compound
classes, and often provide information about the separation and
detection of the compound under examination. Literature is also

published periodically by the companies that supply materials for TLC. For a list of such companies, refer to the appendix at the end of this volume. Company newsletters often publish details of separations not yet available in conventional journals. Furthermore, these brochures may provide the detail necessary for the success of a separation, since journals do try to keep detail to a minimum in many instances.

REFERENCES

1. N. A. Izmailov and M. S. Shraiber, *Farmatsiya*, *3*, 1 (1938).
2. J. E. Meinhard and N. F. Hall, Anal. Chem., *21*, 185 (1949).
3. J. G. Kirchner, J. M. Miller, and G. J. Keller, Anal. Chem., *23*, 420 (1951).
4. J. M. Miller and J. G. Kirchner, Anal. Chem., *24*, 1480 (1952).
5. J. M. Miller and J. G. Kirchner, Anal. Chem., *26*, 2002 (1954).
6. E. Stahl, Chemiker Ztg., *82*, 323 (1958).
7. E. Stahl, Ed., *Thin Layer Chromatography*, Springer-Verlag, Berlin, 1965.
8. J. G. Kirchner, *Thin Layer Chromatography*, Wiley-Interscience, New York, 1967.
9. E. Stahl, Ed., *Thin Layer Chromatography*, Springer-Verlag, New York, 1969.
10. J. C. Touchstone, Ed., *Quantitative Thin Layer Chromatography*, Wiley-Interscience, New York, 1973.
11. J. C. Giddings, *Dynamics of Chromatography*, Marcel Dekker, New York, 1965.
12. Anonymous, J. Chromatog. Sci., *8*, D2 (1970).
13. G. B. Marini-Bettolo, Ed., *Thin Layer Chromatography*, Elsevier, Amsterdam, 1964.
14. J. M. Bobbitt, *Thin Layer Chromatography*, Reinhold, New York, 1963.
15. K. Randerath, *Thin Layer Chromatography*, Academic Press, New York, 1963.
16. K. Macek, Ed., *Pharmaceutical Applications of Thin Layer and Paper Chromatography*, Elsevier, Amsterdam, 1972.
17. G. Zweig and J. Sherma, Eds., *Handbook of Chromatography*, Chemical Rubber, Cleveland, 1972.

CHAPTER 2

Preparation of Thin Layer Plates

2.1 BACKING SUPPORTS FOR SORBENT LAYERS

The primary prerequisites for a support material to hold the chromatographic sorbent are mechanical strength, chemical resistance to solvents and visualization reagents, ability to withstand the temperatures needed for reaction on the plate, uniformity of thickness, and low cost.

Glass plates meet these requirements best and therefore are the most widely used, both for "home-made" and commercial precoated layers. Borosilicate glass is preferred over lesser grades for its chemical and thermal stability. Glass-backed plates must be used when visualization is to be carried out with sulfuric acid, for example. Aluminum and plastic supports are attacked by this corrosive acid.

Flexible plastic sheets, commonly of polyethylene terephthalate, are widely used as supports for commercial plates but are seldom coated in the laboratory by the researcher. Two suppliers produce precoated plates using a glass fiber support impregnated with sorbent; this support is not available for home-made plates, however. More will be said on these in Chapter 3. Heavy aluminum foil is also used for some commercial precoated plates.

The 20×20 cm plate has become the most widely accepted, and most apparatus and accessories have been designed around this size. Larger plates, 20×40 cm, are also available for preparative separations.

Using a normal hardware store glass cutter and a straight edge, it is possible to cut a 20×20 plate into smaller plates such as the widely used 5×20 cm and 10×20 cm. If a clean plate

of glass is used as a counter-top surface, it is possible to cut
an already coated plate, home-made or commercial, by laying it
sorbent surface down on the glass and scribing the back of it with
a straight edge and glass cutter. It can then be picked up and
snapped into the desired size pieces.

2.2   PREPARATION OF GLASS PLATES PRIOR TO COATING

It is important to employ good quality, uniform-thickness plates
(glass or aluminum) for the preparation of dependable home-made
thin layers. If you use glass plates, it is advisable to soak
them at least a day in saturated sodium carbonate solution.
Rinse the plates well with distilled water and dry them. The
next step is most important. Handling the plates by the edges,
carefully and completely wipe them well with a good lipid solvent
such as ether or petroleum ether, using cheesecloth or other
lintless cloth. If they are not ready to be coated immediately
with sorbent, store them horizontally in a plate storage rack
(Figure 2.1) in a clean, fume-free, dust-free environment such as
a cabinet or desiccator, as shown in Figure 2.2. It is advisable
to wipe them with lipid solvent immediately before coating if they
have been in storage a while.

Figure 2.1. Storage rack for drying, storing, and transporting
thin layer plates. Courtesy Kontes.

Figure 2.2.   Storage rack in metal desiccating cabinet for the
storage of thin layer plates.   Courtesy Brinkmann Instruments.

2.3   AVAILABLE SORBENTS FOR PREPARING PLATES

2.3.1   Introduction

There are many experimentally proven sorbent materials suitable
for separating substances when they are coated in a thin layer
[usually 250 microns (250 μ - 0.25 mm)] on a support material.
Most of the well-proven adsorbents are commercially available
from a number of reliable firms, and it is neither necessary nor
economical to prepare them in the laboratory.   The preparation,
including extensive washing of the various sorbents presently
available, is a very time-consuming process.   Furthermore, it is

difficult to exactly reproduce the steps from batch to batch.
The commercial availability of carefully controlled and prepared
substrates precludes this effort on the part of the individual
laboratory. The bulk of the primary supplies of many of the sorbents are manufactured by only a few firms, and most TLC outlets
use these materials. These sorbents or commercially coated plates
are relatively uniform, differing only in method of coating and
sometimes the binder used. Special-purpose sorbents, of course,
may involve further treatment in the final manufacturing process.
      The factors to be considered when choosing a sorbent for TLC
are: type of compound to be separated; visualization technique
to be employed; thickness and stability of the layer desired; and
mobile phase characteristics. When these factors are considered
along with the wide variety of adsorbents available, it is usually possible to choose just the sorbent necessary for the desired separation. Also, as a minor consideration, the nature of
the visualization reagent is a factor. Some layers that contain
organic binders or use organic sorbents are not applicable when
a strong reagent such as sulfuric acid plus heat is required.
Each of these factors will be discussed in later chapters.
      A number of commercial firms supply ready-to-use sorbents,
with and without binders, UV indicators, silver nitrate, etc.
The following pages will describe most of these sorbents, including manufacturer and peculiar characteristics.
      The specifications described in the following sections under
each supplier and sorbent were obtained from the most recent
catalogs and bulletins published by the suppliers at the time of
this writing. Table 2.1 lists the adsorbents available from a
number of suppliers.
      Activation is a process whereby a thin layer plate, either
home-made or commercially prepared, is heated at a given temperature for a period of time to drive off excess water from the sorbent layer. Ridding the layer of water activates it in that
sharper, better separations are often obtained than when a plate
is not activated.

      2.3.2 Binders

Sorbents are prepared with and without binding substances. A
binder changes properties of the sorbent so that it will adhere
to the support well enough not to readily flake off during handling, chromatography, visualization, or documentation. Binding
agents are not always necessary or desirable. Cellulose, a widely used sorbent in TLC, does not usually require a binder; it
inherently has good properties of adhesion and strength. Binders
are undesirable when they interfere with development or visual-

TABLE 2.1 SUPPLIERS OF BULK TLC ADSORBENTS FOR COATING PLATES

	Silica Gel	Silica Gel with AgNO₃	Alumina	Cellulose	Polyamide	Cellulose Ion Exchangers	Sephadex	Kieselguhr	Other Available Sorbents
Analabs, Inc.	X								Magnesium oxide (Anasil)
Applied Science	X	X	X	X		X			
J. T. Baker	X		X	X	X	X		X	Other Inorganics
Bio-Rad	X								Silanized Silica Gel
Brinkmann	X		X	X	X	X		X	Silanized Silica Gel
Camag	X			X					
EM (E. Merck)	X		X	X	X	X		X	
ICN (Woelm)	X	X	X	X	X	X			
Pharmacia							X		
Schleicher & Schuell	X			X					
Supelco	X	X		X		X			
Whatman	X			X		X			

21

ization. Calcium sulfate, a commonly used binder, is slightly
soluble in water, and therefore it is not suggested that layers
employing it as a binder be used with aqueous solvent systems.
Layers employing starch or polymers as the binder (so-called organic binders) should not be used when very high temperature
(>150°) visualization techniques are to be used. They are not
suitable with universal reagents such as sulfuric acid spray.
Starch is recommended only when it does not interfere and when
tough, strong layers are needed.

Some commercially available precoated plates are made with a
small amount of organic polymeric binder, which makes the layer
very hard and resistant to abrasion. This binder will not dissolve in relatively strong polar solvents or in organic solvents,
thus increasing the applications for separation but precluding
some detection procedures.

Calcium sulfate binder concentrations range from about 5% by
weight in some silica gel preparations to 9% in aluminum oxide G
sorbent, 13% in the widely used silica gel G, and up to 30% in
some silica gel preparations for preparative (high capacity) TLC.
The "G" in these names denotes the presence of calcium sulfate
(gypsum).

Adsorbents without a binder are usually designated with an
"H" or an "N" in the name, for example, silica gel H.

2.3.3 UV Indicators

Most commercially available sorbents may be obtained with or
without a chemically inert ultraviolet (UV) indicator for visualization of many compound types. The indicator, at a concentration of 2%, is usually manganese-activated zinc silicate with an
activation peak at 254 nm of light. A thin layer prepared with
the indicator will emit a light green light under a shortwave
(254 nm) UV light. If a substance that absorbs light in this
region is present in the separated zone, it will show up as a
dark area against the green fluorescent background. A long-
wavelength (366 nm) UV indicator, which has a blue fluorescence,
is also available. The presence of one of these indicators in an
adsorbent is designated with an "F", F-254 or F-366. Normally
when just an "F" is used it means the short-wavelength (254 nm)
indicator is present. A sorbent with a UV indicator is useful in
detection procedures only when the separated compounds absorb
light in the region of the wavelength of maximum activation of
the phosphor incorporated in the sorbent, i.e., at 254 or 366 nm.

2.4 SILICA GEL OR SILICIC ACID

Silica gel has become the most widely used sorbent for thin layer chromatography since being described by Kirchner et al. in
1951 (1). The book by Stahl (2) deals extensively with the properties and preparation of silica gels used in TLC.

Analabs, a division of New England Nuclear, offer their own
grade of silica gel under the trade name Anasil. It is available
with 13% calcium sulfate binder (G), with binder and fluorescent
indicator (GF), without binder or indicator (H), and without
binder but with UV indicator (HF). Available package quantities
are 500 g, 1 kg, and 5 kg.

Adsorbosil is the brand name of the silica gel adsorbents
supplied by Applied Science Laboratories. Every batch manufactured is pretested to ensure that it will perform a specific
separation. A quality check certificate supplied with each bottle gives all details of the test separation and directions for
making the slurry. The manufacturer generally recommends mixing
50 g Adsorbosil with 60 ml distilled water, spreading the plates,
then allowing them to air dry. Activation is recommended at 110°
for 30 min before use.

The Adsorbosil line represents a complete offering of silica
gel adsorbents with all combinations of no binder, calcium sulfate, or magnesium silicate binders, and with or without fluorescent indicator.

Also available is another Adsorbosil that is impregnated
with 20% silver nitrate, with or without 10% calcium sulfate
binder. Chromatography on silica gel impregnated with silver
nitrate allows compounds containing double bonds to be separated
on the basis of their cis-trans configuration and degree of unsaturation. Cyclic olefins are readily separated from their
straight-chain counterparts. Normally, the fewer the double
bonds present in the compound, the faster it migrates. Some applications include separations of unsaturated fatty acids (3),
mono- and sesquiterpenes (4), resin acids (5), chlorophylls (6),
and quantitation of pesticides (7). Visualization is often accomplished by charring with potassium dichromate-sulfuric acid.

For convenience, Applied Science Laboratories also offer
silica gel with and without binder, and with and without UV indicator, prepared by E. Merck in Darmstadt, West Germany.

J. T. Baker Chemical Co., Phillipsburg, New Jersey, carries
three forms of silica gel for TLC. Silica gel 7 is a neutral
form containing no binder; silica gel 7A contains calcium sulfate
as a binder. Silica gel 7 GF contains this binder as well as
long- and short-wave UV indicators.

Brinkmann Instruments, Westbury, New York, is one of the
largest suppliers of TLC sorbents, precoated plates, and appa-

ratus. They are distributors for the EM (E. Merck) and Brink-
mann/MN (Macherey, Nagel) lines of sorbents and plates, and the
Desaga line of apparatus. In addition to the usual forms of
silica gel offered by EM and MN, Brinkmann also offers silica gel
specially prepared for making preparative (not analytical) layers. Preparative layers are usually 500-2000 μ (0.5-2.0 mm)
thick as compared to 100 or 250 μ thickness normally used for analytical layers. The thicker preparative layers permit the application of larger sample amounts because of increased capacity.
The primary function of preparative TLC is the isolation of quantities of a substance by employing TLC techniques. When sorbents
that are intended primarily for analytical thin layers are used
to make thicker preparative layers, cracking or pitting of the
layer often results. The specially manufactured "P" (preparative) sorbents were developed to eliminate some of these problems. These "P" silica gels are available in four different preparations as the normal silica gel with and without calcium sulfate binder, and with or without fluorescent indicator. Dual-
wavelength (254+366 nm) indicator is also available with the preparative silica gel.

    Another large supplier of TLC apparatus, plates, and sor-
bents is Camag, Inc., which is affiliated with Camag of Switzer-
land. They offer eight different types of silica gel for layer
preparation, including analytical and preparative grades, binder
or no binder, and indicator or no indicator. Their binder is
usually 5% calcium sulfate. The designation used for the analytical layer silica gel is "D"; D-0 is without binder or indicator; D-5 contains binder only; DF-5 contains both; and DF-0
contains indicator only. The preparative grade silica gel is
designated "DS," and all forms follow the same coding system as
the analytical grades.

    E. Merck adsorbents may be obtained directly from EM Labora-
tories in Elmsford, New York. Their line of silica gel sorbents
are prepared according to Stahl (2), with extra pure (type HR)
and silanized types available along with the usual types available from the other suppliers. They also have silica gel H,
HF-254, and HF-254+366 grades, where the "H" designates that the
binder is composed of fine particles of silicon dioxide or alumina. All preparative silica gel has F-254 indicator, and a
silanized type is also available for reversed-phase chromatography.

    Supelco, Inc., offers eight types of silica gel for TLC.
Their own product trademark is Supelcosil, which is available
with and without UV indicator and with or without binder. They
offer two unique specially prepared forms of silica gel:
Supelcosil 12D, which uses silicon dioxide as a binder and contains 15% silver nitrate, and Aflasil, which has calcium sulfate

binder and is used for the separation of glycolipids and aflatoxins. Whereas most silica gel has a pH of 7.0, Aflasil has a
pH of 7.4-7.6.

2.4.1 Preparation of Silica Gel Slurry

ICN Pharmaceuticals is the distributor of Woelm chromatographic
products in the United States. These include adsorbents, precoated plates, and visualization sprays. They offer the four basic types of analytical layer TLC silica gel, with and without 1%
calcium sulfate binder, and with and without fluorescent indicator.
    For batch spreading of five 20×20 cm plates (or equivalent,
ten 10×20 cm; twenty 5×20 cm) with silica gel G, mix 25 g powder
with 50 ml distilled water. For spreading one 20×20 cm plate by
hand, mix 5 g silica gel G with 15 ml of a 2:1 (v/v) solution of
ethanol:water.
    The quantities are somewhat different when using silica gel
without the binder. For the batch coating of five 20×20 cm
plates using a spreader, mix 30 g powder in 45 ml water. For
pouring one plate by hand without a spreader, mix 6 g powder in
15 ml of a 9:1 ethanol:water solution.
    Woelm also manufactures a silica gel GPF-254 for coating
preparative layers. Their formula for one 20×20 cm plate, 1 mm
thick, is 20 g silica gel GPF mixed with 30-35 ml water. The
plate can be activated at 130° before use.
    Schleicher and Schuell, Inc., well known for filter and
chromatography papers, supplies adsorbents in bulk as well as
precoated TLC plates. Six different forms of silica gel are
available: the four commonly supplied (with and without UV indicator and with or without 12% calcium sulfate binder) and two
unusual ones, with starch (3%) as a binder, with and without UV
indicator. The following procedures are offered for preparation
of these silica gels.
    For silica gel without binder, slurry 30 g powder in 60-65
ml water. Homogenize with an electric blender for 30 sec.
Spread the slurry 250 µ thick on five 20×20 cm plates. Dry at
ambient temperature, then activate at 110° for 30 min.
    For silica gel with calcium sulfate binder, slurry 30 g
powder in 65-70 ml water. Using an electric blender, homogenize
for 30 sec and spread five 20×20 cm plates within 2 min. Dry,
then activate at 110° for 30 min.
    For silica gel with starch binder, slurry 30 g powder in 90
ml boiling water. Homogenize for 30 sec with an electric blender and spread while hot. This recipe will make five 20×20 cm
plates. Dry and activate at 110° for 30 min.

Bio-Rad Laboratories also supplies a silica gel suitable for TLC. It is without binder and is called Bio-Sil A for TLC.

Whatman supplies the Reeve Angel brand of silica gels for TLC. Included is silica gel without binder or phosphor, without binder but with phosphor, with both binder and phosphor, and with binder but without phosphor.

For the preparation of home-made silica gel plates containing silver nitrate, Applied Science (8) recommends making a slurry by stirring 20 g "good quality" TLC silica gel powder into a solution of 10 g silver nitrate dissolved in 50 ml water. Stir the slurry carefully to remove air bubbles; three 20×20 cm plates 250 μ thick may be coated. Allow the plates to air dry for 2 hr in a protected area, and then activate them for 1 hr at 110° before using. Plates may be prepared in advance of their use but should not be activated until ready for use. The prepared plates may turn light gray upon activation, but will fade after developing and visualization by charring with sulfuric acid and heat. Applied Science recommends using its own Plexiglas plate spreader to coat the plates because of the corrosive nature of silver nitrate on the metal used in other spreaders. The coated plates should always be stored in the dark and only for limited periods of time, as $AgNO_3$ layers will gradually darken; the process is hastened by light.

2.5 ALUMINA

Aluminum oxide, commonly called alumina, is the second most widely used TLC sorbent after silica gel. Many applications for alumina in TLC have been adapted from similar applications in column chromatography. As with silica gel, good TLC grade alumina is available from a number of commercial suppliers for making plates in the laboratory. Precoated alumina plates are also readily available; these will be discussed in Chapter 3.

Alumina is manufactured with three pH ranges--acid, neutral, and basic--for separating different types of compounds. To obtain reproducible results with alumina, it is usually necessary to control the amount of adsorbed water, which can block the active separating sites on the alumina surface. To do this, it is necessary to activate the coated TLC plate in an oven at specified temperatures between 75- and 110° for a specified period of time before applying the samples to the plate. All the parameters in the preparation of the plate must be defined in order to reproduce a given separation.

Applied Science Laboratories offers aluminum oxide GA, which contains 10% calcium sulfate binder, and aluminum oxide HA, which has no binder. Both have a pH range of 9.0-9.5 and a low iron

content (less than 0.01%). In the past, iron was often a troublesome contaminant, but with the specially prepared chromatographic
grades of alumina available now, this problem has been kept to a
minimum.

A basic aluminum oxide containing an inorganic fluorescent
indicator that is activated at 254 or 366 nm is available from J.
T. Baker as aluminum oxide 9F.

Brinkmann supplies a complete line of alumina sorbents. Included are acidic, basic, and neutral pH forms with binder but
without UV indicator. These are cataloged as Type T aluminas,
which are suggested for peptide and steroid analyses, among other
uses. Also available are aluminum oxide G (binder), FG (binder
and indicator), and H (no foreign binder). All are suggested for
analytical work.

Alumina suitable for the making of preparative layer plates
is also supplied by Brinkmann. This alumina contains either
short-wave (254 nm) or short-wave and long-wave (366 nm) ultraviolet indicators. Both indicator pigments are available separately and may be added to any sorbent before plate coating if
desired.

Alumina without the calcium sulfate binder is recommended
whenever there is a likelihood that calcium ions would interfere
with the separation or detection. It is also suggested for charring visualization techniques. Without the binder, the layer is
more resistant to the solubilizing effects of polar solvent systems used in development. Generally speaking, bulk alumina contains more impurities than other sorbents, making it difficult to
obtain "clean" layers and reproducible results.

2.5.1 Preparation of Alumina Slurry

The following formulas are suitable for the preparation of alumina slurries. For analytical layers using aluminum oxide G with
binder, mix well 30 g powder with 40 ml distilled water. This
will coat five 20×20 cm plates or twenty 5×20 cm to a depth of
250 μ. Immediately after coating, allow the plates to dry for at
least 30 min at room temperature before activation or use. Activation at 110° for 30 min immediately before use is suggested.
Calcium sulfate binder sets very fast, and it is therefore imperative that the plates be spread within 2 min after preparation of
the slurry.

To use aluminum oxide H (without binder) for analytical layers, mix 30 g with 80-90 ml distilled water. Since there is no
fast-setting binder in this case, there is no urgency involved
with coating, and the slurry may even be stored for future use.
Also because of this, the plates must be allowed to dry for 3-4
hr before activation or use. They should be activated for 1 hr

at 110°.

For preparative plates, it is suggested that the preparative quality aluminum oxide P be used. The slurry preparation is different. The following formula will coat two 20×40 cm preparative size plates, or four 20×20 cm plates to a depth of 2 mm (2000 μ). A proportionate number of plates may be coated to 1 mm (1000 μ) or 0.5 mm (500 μ) if desired. Mix 320 g alumina with about 320 ml distilled water, shaking well for 1 min. An electric blender may also be used. Allow the slurry to stand for 15 min, then shake it again and spread it on the plates. The freshly coated plates must then be stored horizontally for 15 hr at room temperature. Activation and final drying are done at 110-120° for 3-4 hr.

Camag supplies aluminum oxide DS. The "D" designates that it is their own brand, and the "S" designates that it is made up of slightly larger particles than usual so that the separation is faster than normal. Four types are available: aluminum oxide DS-0 has neither binder nor indicator; DSF-5 has short-wave UV indicator and 5% calcium sulfate binder; DS-5 has binder without indicator; and DSF-0 has indicator without binder. Types DS-0 and DSF-0 are recommended for preparative layers of 2 mm and over.

EM Laboratories offer a complete line of alumina sorbents. Two basic types and quite a number of additional forms of each type are offered. Type T is aluminum oxide ignited at a higher temperature than Type E. Brinkmann uses the same designations for their fundamental types. Acidic, basic, and neutral forms of Type T are available for specific pH-dependent separations. Type E aluminum oxide is offered with 10% calcium sulfate binder as aluminum oxide G; with binder and indicator, GF-254; with binder composed of fine particles of silicon dioxide or aluminum oxide, H; with this binder and indicator, HF-254.

Preparative forms of both Type E and Type T are available, but only with UV indicators, either short-wave (254 nm) or short-wave combined with long-wave (254+366 nm).

ICN Pharmaceuticals distributes four different types of Woelm alumina. Alumina G contains 10% calcium sulfate binder is stocked, as are basic, neutral, and acid alumina without binding agent. None of these four is supplied with UV indicator. However, 2% fluorescent indicator green (254 nm, catalog #404-166) can be added to any of the four during preparation of the slurry if fluorescent indicating plates are desired.

For coating plates with these sorbents, mix 35 g with 40 ml distilled water. Using a spreader, spread the alumina G immediately because the binder is fast setting; the other three forms do not have to be spread as quickly. This recipe is sufficient for five 20×20 cm plates. Activate the plates for 30 min at 130°.

If a spreader is not available, the sorbents may also be

spread by pouring, but this is not recommended on a routine basis because of the nonuniform, unknown thickness of the layers produced. A single plate may be coated with alumina G by mixing 6 g with 15 ml of 2:1 ethanol:water. If the other aluminas are used, mix 6 g with 15 ml of 9:1 ethanol:water.

2.6 CELLULOSE

Silica gel and alumina are both inorganic substances that are used chiefly to separate lipophilic ("fat-loving") compounds or those soluble mainly in organic solvents. Cellulose, on the other hand, is organic and is mainly used to separate hydrophilic ("water-loving") compounds such as sugars, amino acids, inorganic ions, and nucleic acid derivatives. These substances are soluble in aqueous solutions. The behavior of cellulose coated in a thin layer on a glass, plastic, or aluminum support is not much different from that of a paper strip in paper chromatography. The developing solvents and visualization reagents employed in paper chromatography are most readily used in cellulose thin layer chromatography. The separation process occurring on cellulose is mainly that of partition because the adsorbed water molecules cover the cellulose molecules, being joined to them by hydrogen bonds. The substances to be separated on a cellulose layer are actually partitioned between the developing solvent and the water coating the cellulose particles. Cellulose thus has many hydrophilic properties because of this adsorbed water, and very little adsorption of sample components actually occurs.

Two types of normal cellulose powder have been found suitable for TLC: native, fibrous cellulose and microcrystalline cellulose. The fibrous cellulose is produced primarily by Macherey, Nagel Co. (MN) under the name MN 300 cellulose. In the United States the distributor for MN products is Brinkmann. The major brand of microcrystalline cellulose is Avicel, made by American Viscose. They supply it to the chromatography companies for sale in bulk or for the manufacture of plates, known either as Avicel plates or microcrystalline cellulose plates, depending on the supplier.

Applications for the two celluloses are not always the same, one providing a better separation than the other for a particular set of compounds. There are six forms of chemically treated cellulose available commercially for TLC use. These forms and their suppliers are listed in Table 2.2. These are ion-exchange sorbents (except acetylated cellulose, which is used for reversed-phase chromatography) and are widely used for separating larger molecules because the active substance is present mostly on the surface of the cellulose, where it can readily interact with the

TABLE 2.2 SUPPLIERS OF CELLULOSES FOR THIN LAYER CHROMATOGRAPHY

	Cellulose, Fibrous "Cellulose 300"	Micro-crystalline*	DEAE†	Ec-teola	PEI‡	Acetylated	Carboxy-methyl (CM)	Phosphoryl-ated	QAE§
Applied Science		X	X	X					
J. T. Baker		X	X	X	X	X	X		
Brinkmann	X	X	X	X	X	X	X	X	
Camag	X	X							
EM (E. Merck)		X	X	X	X	X	X		
ICN (Woelm)	X		X	X	X	X	X	X	X
Schleicher & Schuell	X	X	X	X	X	X	X	X	X
Supelco		X							
Whatman	X	X	X	X			X	X	

* Often called Avicel (American Viscose trade name)
† DEAE = diethylaminoethyl
‡ PEI = polyethylenimine impregnated
§ QAE = quaternarized ion exchanger

molecules to be separated.   These larger molecules include protein, peptides, steroids, and enzymes.

Three of these celluloses are derivatives of cellulose formed through chemical reaction.   They include carboxymethyl (CM)
cellulose, DEAE cellulose, and Ecteola cellulose.   The other two
forms, PEI and phosphorylated, are impregnated cellulose in which
the cellulose behaves primarily as a support for the ion-exchang-
er.   A brief explanation of each of these cellulose forms may be
helpful.

Carboxymethyl (CM) cellulose, R-0 · CH_2COOH, is synthesized
from monochloroacetic acid and alkali cellulose.   It is a weak
cation exchanger.

Diethylaminoethyl (DEAE) cellulose, R-0 · CH_2H_4 · $N(C_2H_5)_2$,
is the reaction product of 2-chloro-1-diethylaminoethyl hydrochloride and alkaline cellulose.   It is a strong basic anion exchanger.

Ecteola cellulose is the product of the reaction of alkaline
cellulose with triethanolamine and epichlorohydrin.   It is a weak
basic anion exchanger.

Polyethylenimine (PEI) cellulose is fibrous cellulose impregnated with polyethylenimine.   It is a strong basic anion exchanger.

Phosphorylated (P) cellulose, R-0 · PO_3H_2, is a strong cation exchanger formed from alkaline cellulose and phosphorous oxy-
chlorine or fibrous cellulose and urea phosphate.

Because cellulose has natural adhesive properties, the addition of a binder does not increase the stability of the layer.
Most bulk cellulose sorbents, as well as precoated plates, are
therefore offered without binder.

The relative popularity of the different forms of cellulose
is shown by the degrees of availability from the commercial suppliers.   Microcrystalline cellulose is the most widely used form
and the most generally available.

Applied Science Laboratories supplies microcrystalline,
DEAE, and Ecteola celluloses.

Microcrystalline cellulose 10% acetylated cellulose, cellulose CM, DEAE, Ecteola, and PEI are available from J. T. Baker.

Brinkmann stocks all eight forms of cellulose from cellulose
300 to the phosphorylated form.   All forms are normally supplied
without UV indicator except for the normal fibrous cellulose 300,
which is available with or without it.   Brinkmann also has specially prepared cellulose 300-HR (MN), which is acid washed and
fat-free, for use when a highly purified cellulose is desired.
Four forms of cellulose 300 are available: 10%, 20%, 30%, and
40%.   The percentage designation is the degree of acetylation of
the cellulose.

Cellulose powder D is the Camag equivalent to cellulose 300.
It is available with and without UV indicator. Microcrystalline
cellulose is also available from Camag with and without UV indicator as cellulose powder DS.

EM Laboratories supplies all forms of cellulose except the
300 and phosphorylated types.

ICN Pharmaceuticals distributes all forms of cellulose for
TLC application except the microcrystalline and acetylated types.
They also offer a special form not readily found elsewhere, quaternarized ion exchanger (QAE). This is an anion exchanger used
for the separation of proteins and other delicate biological substances as well as weakly acidic substances. It is also suitable
for partition TLC in polar solvents.

Schleicher and Schuell supply all nine forms of cellulose
for coating plates: normal acid-washed, microcrystalline, and
20% and 40% acetylated. Also available are DEAE, Ecteola, carboxymethyl cellulose (CM), PEI, QAE, and phosphorylated cellulose.

2.6.1 Preparation of Cellulose Slurry

Preparation of slurries is different for each of the celluloses.
Each of the following formulations will coat five 20×20 cm plates
or the equivalent in smaller plates: ten 10×20 cm or twenty 5×20
cm.

For normal cellulose (200 μ thick), dissolve 0.4 g starch in
10 ml distilled water and add to 90 ml boiling water. Boil for
1 min and add 20 g cellulose. Using an electric blender, homogenize for 30 sec; coat the plates immediately while the suspension
is still hot. Dry in air or for 45 min at 110°. No activation
is necessary if drying occurs at 110°.

To form a slurry of microcrystalline cellulose (without UV
indicator, layer thickness 100 μ), mix 20 g in 60 ml distilled
water. Homogenize for 30 sec using an electric blender. Let the
plates dry in air or at 110° for 30 min. No activation is necessary if plates are dried at 110°. Blending time is critical; the
degree of hydration increases with increased blending time. This
changes the nature of the thin layer and affects the speed and
quality of the separation. This variable should therefore be controlled for reproducible layers.

To prepare a slurry of microcrystalline cellulose with UV
indicator, add 25 g to 40 ml methanol plus 20 ml distilled water.
Homogenize for 30 sec in an electric blender. Dry the spread
plates in air or at 110° for 30 min. No activation before use is
necessary if drying occurs at 110°. The methanol in this formula,
as in the silica gel with UV indicator formulation, is necessary
to solubilize the fluorescent green indicator so that it will be
uniformly distributed throughout the slurry.

A slurry of acetylated cellulose (for layers 130 µ thick) is prepared by mixing 30 g sorbent plus 4.5 g calcium sulfate ($CaSO_4 \cdot \frac{1}{2}H_2O$) in 60 ml distilled water with 10 ml methanol. Blend for 30 sec in an electric blender. Air bubbles may be removed by carefully covering this suspension with 2-3 ml methanol and shaking. Spread the plates within 10 min, and dry them at room temperature or for 30 min at 110°. No activation is necessary if plates have been dried at 110°.

Randerath and Randerath (9), in an excellent paper on coating plastic sheets with PEI cellulose, state that "the coated sheets are allowed to dry overnight at room temperature. This will result in a layer with optimal ion-exchange properties. Under no circumstances should drying of the layer be accelerated by heating." This is the usual case, for most often cellulose is used for partition chromatography, where the atmospheric water retained by the cellulose in the layer is necessary for the proper separations to occur. It appears to be the exception, therefore, to recommend drying a cellulose plate with heat as stated in the previous paragraphs.

Brinkmann suggests different formulas for the preparation of cellulose slurries using the Macherey, Nagel (MN) brand of celluloses. The following formulations will each coat five 20×20 cm plates or the equivalent number of smaller plates.

For MN cellulose 300, mix 15 g with 90-100 ml distilled water in an electric blender for 30-60 sec. Mixing by hand is not aggressive enough for good hydration, and an electric blender is mandatory for cellulose slurry preparation. Other sorbents, e.g., silica gel, alumina, polyamide, and kieselguhr, are just as easily prepared by shaking in an Erlenmeyer flask. The coated cellulose plates are air dried with no activation.

Slurry preparation of MN cellulose 300 with 254 UV indicator requires 15 g in 60 ml 95% methanol with blending for 30-60 sec in an electric blender.

Avicel cellulose (microcrystalline) must be prepared carefully. Mix 15-20 g with 100 ml distilled water and, using an electric blender, slurry for exactly 60 sec. Longer blending will result in an undesirable gel formation. Dry the plates 30 min at room temperature, then at 105° for 10 min to drive off excess water.

MN cellulose 300 acetate is prepared by first forming a stiff paste from 15 g powder and a small amount of 95% ethanol taken from a total volume of 60 ml. After the paste is formed, add the remaining alcohol and slurry in an electric blender for 30-60 sec. Air dry only after spreading the layers.

Carboxymethyl cellulose, from MN cellulose 300 CM, is prepared from 15 g powder and 60-70 ml distilled water. Blend for 30-60 sec in an electric blender. If dry layers made from this

formulation crack, an alternative formulation may be tried. Mix
12 g with 3 g regular MN cellulose 300 in the same amount of
water. Blend, spread, and dry in the same manner.
   DEAE cellulose MN 300 slurry is prepared from 10 g powder
and 75 ml distilled water. The formulation recommended if this
one produces layers that crack is 8 g DEAE powder plus 2 g cellulose 300 powder in 75 ml water. Blend either formula for 30-60
sec with an electric blender. Air dry after spreading on the
support.
   MN cellulose 300 Ecteola slurry is prepared from 10 g cellulose and 50 ml distilled water. The alternative formulation
used against cracking is 8 g Ecteola plus 2 g cellulose 300 with
50 ml water. Slurry for 30-60 sec in an electric blender. Allow
the plates to air dry after spreading.
   Phosphorylated MN cellulose 300 P is slurried from 15 g in
60-70 ml distilled water. An optional formulation to decrease the
likelihood of cracking is made from 12 g cellulose P plus 3 g cellulose 300. Blend for 30-60 sec in an electric blender. Allow
to air dry after coating on the support.
   Predevelopment of cellulose plates in a tank containing distilled water is recommended to remove impurities normally found
in most celluloses. This washing will concentrate the impurities
along the upper edge of the plate out of the separation zone.
Then apply the substances to be separated on the "clean" end of
the plate, so that the impurities will not be redistributed over
the plate by the developing solvent. Washed plates can be marked
with an arrow or other notation using a scribe, e.g., syringe
needle, on the upper or "dirty" edge.

       2.7 POLYAMIDE

On polyamide thin layers, separation depends on the strength of
the hydrogen bonds formed between the polyamide molecules and the
substances being separated. This bond strength is a function of
the number of phenolic hydroxyl or carboxyl groups and their positions in the molecules to be separated. Compound types that
can be separated on polyamide therefore include phenols, carboxylic acids, steroids, quinones, and aromatic nitro compounds. A
monocarboxylic acid, having only one carboxyl group, is not as
readily bound to the polyamide molecule as would be a dicarboxylic acid or an aromatic acid molecule, and will therefore
migrate further on a polyamide thin layer than either of the other
two molecular types. For elution, a solvent must be chosen that
is able to break the hydrogen bonds formed between the sample
molecules and the polyamide molecules by displacement. Certain
solvents are better at doing this than others. Chapters 5 and 6

on solvent systems and development will explore this subject.
There are four different polyamides currently being used for
TLC.
Polyamide 6 = Nylon 6 = aminopolycaprolactam
Polyamide 6.6 = Nylon 6.6 = polyhexamethyldiaminoadipate
Polyamide 11 = Nylon 11 = polyaminoundecanoic acid
Acetylated polyamide = acetylated derivatives of the above
Only four suppliers carry polyamide in bulk for coating
plates. Baker has polyamides 6 and 11. Brinkmann distributes
the polyamides made by E. Merck (EM brand) and Macherey, Nagel
(MN brand). EM Laboratories supplies E. Merck polyamide and ICN
Pharmaceuticals has Woelm polyamide.
Brinkmann supplies E. Merck polyamide powder 11 without UV
indicator, MN polyamide 6, MN polyamide 6 with UV 254 indicator,
and acetylated polyamide 6. Also available are MN polyamide 6.6,
MN polyamide 6.6 with UV 254 indicator, and acetylated polyamide
6.6.
EM Laboratories supplies E. Merck polyamide powder 11 with-
out UV indicator.
ICN carries a "polyamide TLC."

2.7.1 Preparation of Polyamide Slurry

The polyamides commercially available for TLC are normally supplied without a binder, because it is not necessary. When preparing polyamide slurries you must use an electric blender for
thorough mixing. Hand shaking in an Erlenmeyer will not adequately prepare a polyamide slurry.
Macherey, Nagel (MN) brand polyamides, available from Brinkmann, are prepared for coating according to the following procedures. For a polyamide 6 slurry, add 15 g powder to 65 ml distilled
water. Mix well in an electric blender for 30-60 sec and spread
the slurry on the support immediately before any separation of
liquid and solid occurs. Air dry after coating the support.
Polyamide 6.6 is prepared in exactly the same manner.
For a slurry of polyamide 11, blend 15 g powder with 60 ml
methanol in an electric blender for 30-60 sec. Spread the plates
immediately and allow them to air dry.
It is not necessary to activate polyamide layers before use.
Occasionally, if a stable layer is necessary, 10% starch can be
added as a binder.
Rosler et al. (10) developed a standardized procedure for
the preparation of TLC quality polyamide.

2.8 KIESELGUHR

Kieselguhr, known also as diatomite or diatomaceous earth, is
composed of the siliceous skeletal remains of microscopic marine
animals called diatoms. Massive accumulations of diatomite occur in many parts of the world. Because of its high porosity and
large surface area, it is widely employed as a filter aid in the
laboratory, in industry, and in swimming pool filtration. It is
used in column chromatography and as a sorbent in TLC, where it
is often coated with a liquid phase and used for partition chromatography.

When used in TLC, it normally has a binder of calcium sulfate and is known therefore as kieselguhr G, the "G" being the
designation for the calcium binder, usually with a concentration
of about 15%.

Brinkmann and EM are both suppliers of E. Merck kieselguhr
G. These are the only two sources of purified kieselguhr for TLC.

A slurry is prepared from 30 g powder in 60 ml of distilled
water. The plates should be coated within 2 min because the
binder sets rapidly. The layers are allowed to dry at least 30
min at room temperature before activation. Activate for 30 min
at 110°. Kieselguhr is not normally used for preparative TLC.

Kieselguhr G or N (no binder) is prepared from 20 g powder
in 55 ml distilled water. Shake in an Erlenmeyer flask and
spread the G sorbent on the support within 2 min. The kieselguhr
G is allowed to dry at least 30 min, the kieselguhr N at least 4
hr, before activation at 110° for 30 min.

2.9 SEPHADEX

Sephadex is the trade name for a group of modified dextran gels
for gel filtration that are produced by Pharmacia Fine Chemicals.
Sephadex gels are hydrophilic, neutral stationary phases that may
be used in thin layers. The American distributor is Pharmacia in
Piscataway, New Jersey.

Separation in gel filtration is achieved according to molecular size on a gel swollen in a carefully prepared solution.
The mechanism of separation is primarily one of partition governed by steric hindrance. Determann (11) and Johansson and Rymo
(12) were the first to use the thin layer gel filtration technique (TLG). It is basically the same as TLC but differs from it
in that it is carried out on a prewetted and wet layer with the
movement of liquid being continuous. Solvent flow is potentiated
by gravity. Development of Sephadex plates will be covered in
Chapter 6.

As in TLC, plates to be coated with Sephadex should be thoroughly cleaned by washing with warm detergent solution and distilled water. Washed plates may be stored in concentrated sodium
carbonate solution, then washed with distilled water and dried
before coating.

Most of the Sephadex grades are available in a variety of
mesh sizes. For TLG (thin layer gel) chromatography, the "superfine grade" should be used. The consistency of the slurry is important. If it is too thick, the gel will not adhere. If it is
too thin, it will run on the plate. A binder is not normally
used with Sephadex because of its natural adhesive properties.
Table 2.3 provides the information needed to prepare slurries of
each of the Sephadex types. It is necessary to swell the gel in
the eluent developing solvent to be used for at least the amount
of time shown in the table. Air bubbles may be removed from the
slurry by placing the container in a boiling water bath. The
slurry should be carefully stirred by hand to completely suspend
all the Sephadex and break up all clumps. Mechanical and magnetic stirrers should not be used, as they may cause destruction
of the particles.

The gel slurry may be spread by hand or with any commercially available spreader. Pharmacia offers their own spreader,
which is simple in design and easy to use. It is recommended for
spreading Sephadex. Its use is illustrated in Figure 2.3. This
TLG spreader, as it is known, produces even layers of reproducible thickness, which include 0.4, 0.6, 0.8, and 1.0 mm. The
thickness of 0.6 mm (600 μ) is recommended for most uses.

2.10 COATING PROCEDURES

There are four methods that may be used to coat a plate with a
thin layer of sorbent. These are: (1) pouring, (2) spreading,
(3) immersing, and (4) spraying. Of these four, the spreading
procedures are the most widely used, the others being occasionally used under particular circumstances. For instance, microslide
plates are often prepared by the immersion technique because of
the ease with which it is done with small plates. Spreading procedures are preferred for a number of reasons, including good reproducibility, adjustment of layer thickness, and ease of operation. The other three techniques generally lack these virtues.
Spraying techniques are messy and require large guns and high
pressure, which are not ideal laboratory equipment. The spreading techniques, therefore, will be the only coating procedures
covered in the present discussion.

The object in plate coating is to spread a smooth, uniform
layer, which dries without bubbles, pits, or cracks. All of

TABLE 2.3 PREPARATION OF SEPHADEX SLURRIES FOR TLC

Type of Sephadex Superfine	Gel wt. g/100 ml	Minimum Swelling Time	
		At Room Temp.	On Boiling Water Bath
G-50	10.5	3 hr	1 hr
G-75	7.5	24 hr	2 hr
G-100	6.5	72 hr	3 hr
G-150	4.8	72 hr	3 hr
G-200	4.5	72 hr	3 hr

Layer Thickness	GEL SLURRY VOLUME (ml) FOR COATING ONE PLATE	
	20×20 cm	20×40 cm
1.0 mm (1000 μ)	55	110
0.8 mm (800 μ)	45	90
0.6 mm (600 μ)	35	70
0.4 mm (400 μ)	25	50

these defects can cause poor migration of substances to be separated due to "channelling" or trough effects. Three important steps should be taken in plate preparation to help prevent this. Although all have been mentioned previously, they are emphasized here. First, using a lint-free cloth (cheesecloth), wipe the plates well with a fat solvent such as ether or petroleum ether before coating. The sorbent will not adhere to a dirty, greasy support. Careful slurry preparation is the second point to consider. Good quality solvents should be used to prepare the slurry, including deionized and distilled water.

The slurry should be stirred with care so that all lumps of sorbent powder are broken up and no air bubbles remain. After careful coating, the layer should be allowed to air dry at room temperature in the horizontal position for at least the recommended time. Preparative layers normally require 15-18 hr drying time at room temperature. Do not place freshly coated plates in a hot oven to dry. This will result in drying which is too fast and too severe and will invariably produce a cracked layer, often with cracks not easily seen with the eye.

The instructions for slurry preparation for each sorbent given in the previous sections should be followed, as these "recipes" have proven to be reliable.

2.11   SPREADING THE LAYERS BY MANUAL PROCEDURES

The easiest way to coat a plate with a thin layer of sorbent is to simply pour the slurry on top of the plate and spread it evenly across the plate with a glass rod. This will not produce a layer of uniform thickness nor one that can be reproduced unless some provision is made to space the rod at a given height, for example, 0.25 mm (250 μ), above the plate surface so that the spread layer will be uniform. Spacers can be provided on the ends of the rod, as is the case of the Pharmacia TLG spreader (Figure 2.3), or a spacer of the desired height may be placed on either side of the plate to support the spreading rod. Often spacers of this type are made with cellophane or masking tape.

The technique for coating a plate with the Pharmacia spreader rod is illustrated in Figure 2.3. After preparing the necessary amount of gel (a) pour it onto the middle of the precleaned plate and spread around with a spatula; (b) rotate the spreader rod in the slurry until it is evenly wetted; (c) place the rod behind the bulk of the slurry with the Pharmacia spreader (the number facing upwards on the end collar corresponds to the layer thickness); and (d) using the rod, draw the slurry slowly and evenly toward the end of the plate, maintaining an excess in front of the rod. Finish with one continuous, even stroke, tak-

ing excess slurry off the plate. It is convenient to place a
clean glass plate at the end of the plate being coated so that
any excess slurry ends up here rather than on the bench top. If
the plates coated by this procedure are Sephadex gel, they are
then placed in the special developing chamber used for gels to
keep them wet until use. If the plates coated are of the normal
type, then they would be allowed to dry at room temperature for
the minimum length of time before activation or use.

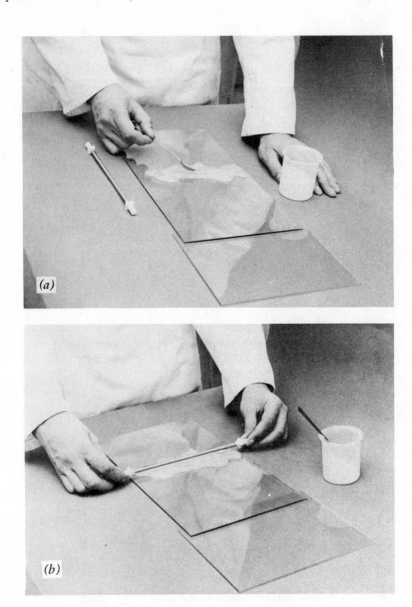

Figure 2.3. Technique of hand spreading a TLC plate using the rod apparatus. Courtesy Pharmacia Chemicals.

2.12 COMMERCIAL SPREADING APPARATUS

There are two basic types of commercially available thin layer
spreading equipment. One type has a fixed trough into which the
slurry is placed, and the plates to be coated are moved underneath. A manually operated form of this apparatus is pictured in
Figure 2.4; the motorized form is shown in Figure 2.5. The second type employs a movable trough containing the slurry, which is
pulled or driven over the top of the plates to be coated. One
such type is pictured in Figure 2.6.

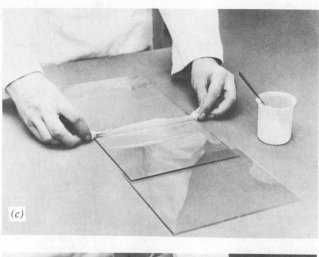

(a)

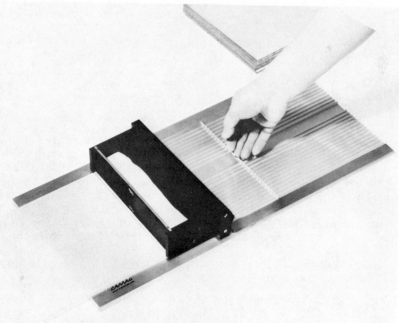

(b)

(c)

Figure 2.4. Procedure for spreading thin layer plates using a
fixed trough apparatus. (a) Clean plates to be coated are placed
on the mounting board, with the edge of the first plate lined up
with the edge of the trough. Slurry is evenly poured into
trough. (b) Plates are slowly pushed under trough at a constant
rate of speed. Do not stop in the middle of a plate. (c) The
finished plate may be taken off the end of the mounting board before the next is pushed through. Courtesy of Camag, Inc.

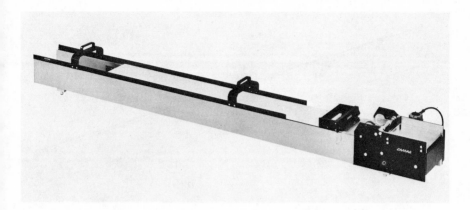

Figure 2.5. A motorized form of the fixed trough spreading apparatus. The finished plates are to the left. The plates to be coated are stacked on the right. A motor-driven wheel pushes the plates under the slurry trough to be coated. Courtesy of Camag, Inc.

(a)

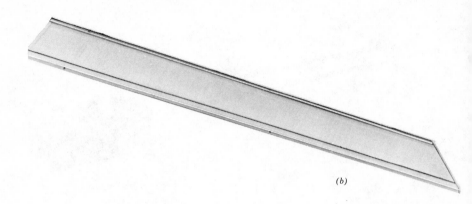

(b)

Figure 2.6. The two major components of a movable trough TLC
spreader. (a) The trough. This one is made of Plexiglas so that
it will not corrode if used with silver nitrate-impregnated silica gel or other corrosive sorbents. (b) The mounting board, or
plate support. This one is made of Plexiglas. The trough slides
in grooves along the outside edges. Courtesy Applied Science
Laboratories, Inc.

     The following points will be helpful when it is desired to
layer "home-made" plates. Set up the apparatus on top of a
chemically clean, dust-free, uncluttered, sturdy, and level bench
or table, preferably where there are no fumes that may be adsorbed into the finished plates while they are drying. Carefully wipe
the plates to be coated with a lint-free cloth soaked with a good
lipid solvent such as petroleum ether to remove any traces of
grease or dirt residue. Set them on the mounting board of the
spreader against the retaining edge, which should be closest to
the operator. Check the plates to make sure that they are all
the same thickness. Do not place plates of different thicknesses
on the board to be coated at the same time. Bumps of sorbent
would be produced, and the thinner plates would receive a thicker
layer of sorbent than the thicker plates because the trough is set
to deposit a given layer depth above the plate surface.
     Often when a movable trough spreader is used, narrow plates
are placed on either end of the plates to be coated to serve as
"start" and "finish" plates to support the trough and slurry so
that all the plates coated are uniform. An end plate is always
advisable to catch any leftover slurry so that it does not flow
over the mounting board or counter.

Microscope slides may also be coated if they are placed close together on a large 20×20 or 20×40 cm plate. If they move about too readily when in contact with the trough, a drop of water under each slide in contact with the supporting glass plate should help them adhere.

The use of the fixed trough spreader is illustrated in Figure 2.4. In (a) the clean plates have been placed on the mounting board and the first plate has been lined up with the edge of the trough. The slurry has been evenly poured into the trough, all ready to spread. In (b) the first plate has been slowly pushed under the trough and has been coated with slurry. Photograph (c) shows the completed plate left on the mounting board so that it may be allowed to dry before the next plate is pushed through. The length of the mounting board is often extended with plastic, plywood, or glass, to support the coated plates as they come out from under the trough and allow room for the next plate. In this way a number of plates may be coated in a continuous motion without interruption or without disruption of the still wet plates.

Operation of the movable trough applicator is not much different than for the fixed trough. All general steps are followed except that the trough is pushed slowly and evenly across the top surfaces of the plates.

Always refer to the instructions supplied with the commercial applicator being used.

REFERENCES

1.   J. G. Kirchner, J. M. Miller, and G. J. Keller, Anal. Chem., 23, 420 (1951).
2.   H. W. Kohlschutter and K. Unger, in Thin-Layer Chromatography, E. Stahl, Ed., Springer Verlag, New York, 1969, p. 7.
3.   D. S. Sgoutas and F. A. Kummerow, Biochemistry, 3, 406 (1964).
4.   M. von Schantz, S. Juvonen, and R. Hemming, J. Chromatogr. 20, 618 (1965).
5.   D. F. Zinkel and J. W. Rowe, J. Chromatogr., 13, 74 (1964).
6.   J. W. Copius-Peereboom and H. W. Beekes, J. Chromatogr., 17, 99 (1965).
7.   M. F. Dobbins and J. C. Touchstone, in Quantitative Thin Layer Chromatography, J. C. Touchstone, Ed., Wiley-Interscience, New York, 1973, p. 295.
8.   Applied Science Laboratories, "Thin Layer Chromatography with Silver Nitrate," Technical Bulletin No. 4 (1964).
9.   K. Randerath and E. Randerath, J. Chromatogr., 22, 110 (1966).
10. H. Rosler, W. Heinrich, and T. J. Mabry, J. Chromatogr., 87, 433 (1973).

11. H. Determann, Experientia, *18*, 430 (1962).
12. B. G. Johansson and L. Rymo, Acta Chem. Scand., *16*, 2067
     (1962).

CHAPTER 3

Commercial Precoated Plates

3.1 INTRODUCTION

Commercially available precoated plates for thin layer chromatography have only been available since about 1967. They are supplied with all the basic sorbents on glass, plastic, or aluminum supports. Sorbents with and without binder and with and without UV indicator are available in a variety of layer thicknesses from 100 μ on most plastic plates up to 2000 μ for some preparative thin layer glass plates. The largest selection is glass, coated to the 250 μ thickness of "analytical layers."

Precoated plates have become quite popular for a number of justifiable reasons. First of all, they eliminate the expensive equipment necessary for the coating process and the trouble and mess that often goes with the procedure. Directly associated with this is the saving of the time needed to prepare a series of plates, which are often only done five (20×20) at a time. This can be a very real saving when a large number of plates and/or a wide variety of sorbents are used, and when labor is in short supply and high priced. Five to six feet of clean, clear bench space is also required to coat a batch of plates, and equipment is necessary for storage. Precoated plates are usually supplied in divided boxes or individually wrapped for convenient shelf storage, which often eliminates the need for even a storage/carrying rack. Convenience is the second major point in favor of the precoated plate.

The most important considerations for their use are that the layers are uniform in thickness and surface, and from package to package the plates yield reproducible results. Both of these features are difficult to duplicate in the laboratory and are of

prime concern in analytical work. High quality commercial precoated plates have been one reason why quantitative thin layer
chromatography has been successful in many applications (1).
    Precoated plates are available with any sorbent except
Sephadex. This allows the chromatographer complete freedom to use
the sorbent needed for a particular application as a precoated
plate and to conveniently stock as many different ones as desired
right on the laboratory shelf.
    Precoated plates with a plastic support have an additional
feature not possible with glass-backed plates: they may be cut
with scissors. This is advantageous when smaller plates are desired but only 20×20 cm plates are on hand. When plastic-backed
plates are used, it is also easy to elute substances out of the
layer by cutting out the area to be eluted with a pair of scissors
and immersing the cut piece in a suitable solvent for extraction.
This is an important consideration when working with radioactive
substances. Counting the radioactivity of the substance is facilitated by simply immersing the piece of plate containing the
radioactive zone in the scintillation vial containing fluid. Because the layer is not scraped; contamination and losses caused
by flying dust are minimized.
    Considering only the cost of the sorbent and the plates, the
precoated plates are more expensive than home-made plates. However, in looking at the total expense involved, the cost of the
coating apparatus, the drying/storage racks, and the labor/time
factor to make the plates must be considered in addition to the
cost of the sorbents and plates when making home-made layers.
After this is done, it will be seen that the apparently high cost
of the precoated plates is indeed reasonable.

        3.2 SUPPLIERS

There are a number of commercial suppliers and manufacturers of
precoated plates, but they do not all carry a complete selection
of sorbent/backing plate combinations in the four major sizes:
5×20, 10×20, 20×20, and 20×40 cm. Some supply the 20×20 cm
plates with prescored lines on the back at 5, 10, and 15 cm, so
that they may easily be broken into the smaller plates. Not all
suppliers provide preparative layer precoated plates because of
their limited, specialized use.
    Four unusual forms of ready-made silica gel and silicic acid
plates are manufactured by Gelman Instrument Co., as a part of
their Instant Thin Layer Chromatography (ITLC) line. The sorbent
is not actually coated on the support but is impregnated in a
glass fiber sheet that serves as the support. These silica gel

plates are produced by immersing very pure glass microfiber sheets into a supersaturated solution of potassium silicate in ammonium chloride. The precipitated silicic acid that is formed surrounds the glass fibers in a hydrated matrix. The ammonia by-product is removed by chromatography in distilled water and by heating at over 300° for a period of time. This silica gel is a weak, very porous stationary phase with a pH of 5 and is suitable for the separation of nonpolar compounds.

Silicic acid plates are prepared in much the same way as silica gel plates, but they contain a greater amount of silica in the form of polysilicic acid. This is denser than silica gel and produces a stronger attractive phase suitable for the separation of polar compounds. The pH of these plates is about 6. A list of the plates available is given in Table 3.15.

Quantum Company Linear-Q divided plates incorporate a 3 cm wide strip as a spotting and preadsorbent sample purification area. Samples and standards are diffusely applied as dilute solutions to the preadsorbent strip in each lane. As development occurs, the compounds travel with the solvent front to the preadsorbent-silica junction, where they are metered into the silica gel as a compact band for separation.

One important factor should be considered when using commercially supplied precoated plates: plates with the same designation from two different manufacturers do not necessarily exhibit the same chromatographic behavior. Given a particular substance or extract to be separated, a precoated silica gel plate from manufacturer A will not separate these substances in exactly the same manner as a silica gel plate from manufacturer B. All other factors remaining the same, even a given type of plate from the same supplier may exhibit variations (2). This is important to keep in mind when trying to repeat a given separation as described in the literature; always use plates of the brand specified in the article if the same results are desired. This is an extremely important consideration in quantitative analysis by *in situ* densitometry, because plates from different manufacturers have different layer characteristics. Even glass and plastic support thicknesses vary from manufacturer to manufacturer. Once a particular separation that works well is achieved, it is important to "standardize" on the plate used; specify and use exactly the same type and brand when repeating the separation.

There is not much literature comparing plate layer characteristics and separations. To the benefit of the method, there are only two or three major manufacturers for each sorbent type, and therefore the amount of diverseness for each sorbent is at a minimum. However, manufacturers employ different coating techniques, thereby changing the characteristics of the sorbent coat-

ed on the plate. So even though two precoated plate companies buy
their sorbent from the same manufacturer, the resulting plates may
not have identical separation characteristics.

Quantum Industries, a manufacturer of precoated plates and
TLC supplies, has prepared a graph that illustrates the comparative polarities of the major brands of silica gel for TLC. This
graph is presented in Figure 3.1. It may be noted that the hard-
ness of a layer will vary according to the brand of silica gel
used, and this in turn affects the speed of development and the
resulting R_f for a given substance. Factors such as activation
and developing chamber saturation are important influences in this
regard also. Generalizations about silica gel layers are difficult to make.

Generalizations about cellulose layers are a little easier
to make. The two most popular forms of cellulose are fibrous and
microcrystalline. The major supplier of fibrous cellulose is
Macherey, Nagel and Co., whose main product is MN 300 cellulose.
The major supplier of microcrystalline cellulose is FMC Corporation, under the trade name Avicel. Microcrystalline cellulose
layers usually develop faster and produce zones that are lower in
R_f and more compact in size than fibrous cellulose layers.

In the following tables layer designations are those used by
the particular manufacturer. The organic binder descriptions are
not specific, because in most cases this information is of a pro-
prietary nature.

Precoated commercial plates employing organic binders do not
adsorb as much moisture from the atmosphere as those made with
calcium sulfate binder and therefore do not usually require activation prior to use.

The addresses of all suppliers are listed in the appendix at
the end of this book.

REFERENCES

1. J. C. Touchstone, Ed., *Quantitative Thin Layer Chromatography*,
    Wiley-Interscience, New York, 1973.
2. P. Turano and W. J. Turner, J. Chromatogr., *90*, 388 (1974).

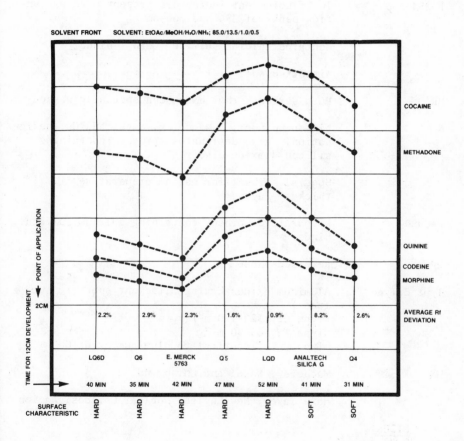

SOLVENT FRONT   SOLVENT: EtOAc/MeOH/H₂O/NH₃; 85.0/13.5/1.0/0.5

Figure 3.1.   Comparative polarities of silica gel TLC plates.
Solvent: EtOAc:MeOH:H₂O:NH₃;85.0:13.5:1.0:0.5.   Courtesy Quantum
Industries.

53

TABLE 3.1   SORBENT SUFFIX DESIGNATIONS

F or F-254	Indicates presence of UV indicator with activation peak 254 nm (short-wave)
F-254+366	Two fluorescent indicators present with activation peaks of 254 and 366 nm
G	Contains calcium sulfate ($CaSO_4$) binder
H	Without binder
HR	Pure sorbents that have been specially washed
N	Without binder, a Macherey, Nagel and Co. designation. Also designates sorbents containing silicon dioxide binder
P	Special sorbent for making preparative (i.e., thicker) layers
P+$CaSO_4$	Preparative sorbent containing calcium sulfate binder
RP	Reversed-phase
Type E/Type T	Aluminum oxides having different specific surfaces: Type E: 120-180 m^2/g Type T: 60-90 m^2/g Type E is identical to aluminum oxide G
MN	Macherey, Nagel and Co. brand
0	Hard layer, New England Nuclear Co. designation

SIZES
a = 20×20 cm
b = 10×20 cm
c = 5×20 cm
d = 15×20 cm
e = 20×40 cm
m = microscope slide, 2.5×7.5 cm
f = 5×10 cm
g = 10×10 cm

TABLE 3.2 SILICA GEL PRECOATED PLATES, GLASS SUPPORT, ANALYTICAL AND PREPARATIVE

Manufacturer	Product	Layer Thickness (μ)	Binder	UV Indicator	Plate Size
Analabs, Inc.	Anasil G	250, 500, 1000	13% $CaSO_4$	-	a, c / a
	Anasil GF	250, 500, 1000	13% $CaSO_4$	+	a, c / a
	Anasil H	250, 500, 1000	--	-	a, c / a
	Anasil HF	250, 500, 1000	--	+	a, c / a
	Anasil 0	250	organic	-	a, b, c
	Anasil 0F	250	organic	+	a, b, c
Analtech, Inc. E. Merck silica gel used	Silica gel G	250, 500, 1000, 1500, 2000	13% $CaSO_4$	-	a-e
	Silica gel GF	250, 500, 1000, 1500, 2000	13% $CaSO_4$	+	a-e; c not available with 1000, 2000 μ
	Silica gel H	250	--	-	a, b, c
	Silica gel HF	250	--	+	a, b, c
	Silica gel HR	250	-	-	a, b, c
	Silicar 7G	250	13% $CaSO_4$	-	a, b, c
	Silicar 7GF	250	13% $CaSO_4$	+	a, b, c

TABLE 3.2 (continued)

Manufacturer	Adsorbent	Amount (g)	Binder	Fluorescent indicator	Remarks
	Silica gel GHR	250	13% CaSO$_4$	−	a, b, c
	Silica gel GHR/UV	250	13% CaSO$_4$	+	a, b, c
	Silica gel G--Woelm	250	13% CaSO$_4$	−	a, b, c
	Silica gel GF--Woelm	250	13% CaSO$_4$	+	a, b, c
	Silica gel G--AgNO$_4$	250	13% CaSO$_4$	−	a, b, c
	Silica gel GF	250	13% CaSO$_4$	+	a, b, c
Applied Science	Silica gel G	250, 500	13% CaSO$_4$	−	a, b, c
	Silica gel GF	250	13% CaSO$_4$	+	a, c
	Silica gel H	250, 500	—	−	a, c
	Silica gel HF	250	—	+	a, c
	Adsorbosil-1	250, 500	10% CaSO$_4$	−	a
	Adsorbosil-1-P	250	10% CaSO$_4$	−	a, c
	Adsorbosil-5	250	—	−	a, c
	Adsorbosil-5-P	250	—	+	a, c
Brinkmann	Silica gel G	250, 500, 1000, 2000	13% CaSO$_4$	−	a-d
	Silica gel G	250, 500, 1000, 2000	13% CaSO$_4$	+	a-d
	Silica gel GHR	250	13% CaSO$_4$	−	a, e
	Silica gel GHR/UV	250	13% CaSO$_4$	+	a, e
	Silica gel/ kieselguhr/UV-254	250	13% CaSO$_4$	+	a

Sorbent	Layer thickness	Binder	Fluorescent indicator	Notes
Silica gel/cellulose/UV-254	250	--	+	a
Camag, Inc.				
Camag silica gel D-B	250	$CaSO_4$	-	a
Camag silica gel DF-B	250	$CaSO_4$	+	a, b, c
Merck silica gel G	250, 500, 1000	13% $CaSO_4$	-	a
Merck silica gel GF	250, 500, 1000	13% $CaSO_4$	+	a, b, c
S&S silica gel	250	organic	-	a, b, c
S&S silica gel F	250	organic	+	a, b, c
EM Laboratory (E. Merck)				
Silica gel 60	250, 2000	organic	-	a, e
Silica gel 60 F-254	250, 500, 2000	organic	+	a
Silica gel 60 silanized RP-2	250		-	a
Silica gel 60 silanized RP-2 F-254	250		+	a
Silica gel 60-kieselguhr F-254	250	13% $CaSO_4$	+	a
High-performance silica gel 60	200	13% $CaSO_4$	-	b, g
silica gel 60 F-254	200	13% $CaSO_4$	+	b, g
ICN Pharmaceuticals (Woelm)				
Silica gel	250	organic	-	a, c
Silica gel F-254	250, 500	organic	+	a, c
Silica gel F-254/366	250	organic	+	a

57

TABLE 3.2 (continued)

Silica gel F-254 "Rapid"	250	organic	+	a, c
New England Nuclear (all pre-fixed with "NEN")				
Silica gel G	250, 500, 1000	13% CaSO$_4$	-	a, c a
Silica gel GF	250, 500, 1000	13% CaSO$_4$	+	a, c a
Silica gel H	250, 500, 1000	- -	-	a, c a
Silica gel HF	250, 500, 1000	- -	+	a, c a
Silica gel 0	250	organic	-	a, b, c
Silica gel OF	250		+	a, b, c
Quantum				
Silica gel Q6 hard surface, very fast	250	organic	-	a-c, f a a
Silica gel PQ6	500, 1000			
Silica gel Q6F	250	organic	+	a-c, f a a
Silica gel PQ6F	500, 1000			
Silica gel LQ6 (has preadsorbent area for rapid application of sample)		organic	-	a, c
Silica gel LQ6F	250		+	a, c
Silica gel LQ6D (silica gel is prescored)	250		-	a, c

58

Layer	Amount (µm)	Binder	Fluorescent indicator	References
Silica gel LQ6DF	250		+	a, c
Silica gel Q5 moderately hard layer	250	organic	–	a–c, f
Silica gel PQ5	500		–	a, e
	1000			a, e
Silica gel Q5F	250		+	a–c, f
Silica gel PQ5F	500		+	a, e
	1000			a, e
Silica gel LQ (Q5 plates with pre-adsorbent for sample application)	250	organic	–	a, c
Silica gel PLQ	1000		+	a
Silica gel LQF	250		–	a, c
Silica gel PLQF	1000		+	a
Silica gel LQD (scored layer)	250		–	a, c
Silica gel LQDF (scored layer)	250		+	a, c
Silica gel Q4 (soft layer)	250	$CaSO_4$	–	a, b, c
Silica gel PQ4 (soft layer)	500			a
	1000			a
	2000			a
Silica gel Q4F (soft layer)	250	$CaSO_4$	+	a, b, c
Silica gel PQ4F (soft layer)	500			a
	1000			a
	2000			a

TABLE 3.2 (continued)

Silica gel Q1 (low polarity, soft layer)	250	organic	–	a-c, e, f
Silica gel PQ1 (soft layer)	500			a, e
	1000			a, e
Silica gel MQ1 (microscope slide)	200			m
Silica gel Q1F (soft layer)	250		+	a-c, e, f
Silica gel PQ1F (soft layer)	500			a, e
	1000			a, e
Silica gel MQ1-F (microscope slide with indicator)	200			m
Supelco, Inc.				
Redi-coat G (silica gel G)	500	$CaSO_4$	–	a, c
Redi-coat AG (silica gel G with 15% silver nitrate)	500	$CaSO_4$	–	a
Redi-coat H (silica gel H)	500	SiO_2	–	a
Redi-coat H (silica gel H with phosphor)	500	SiO_2	+	a, c

Redi-coat (silica gel with magnesium acetate for two-dimensional TLC of lipids)	500	--	-	a
L/S Redi-coat (silica gel with ammonium sulfate for lecithin/sphingomyelin ratio separation	500	--	-	a
Aflasil (specifically prepared silica gel for aflatoxin separation)	500	CaSO$_4$	-	a
Schleicher and Schuell	Silica gel	250 500 1000	organic	-	a a a
	Silica gel	250 500 1000	organic	+	a a a

61

TABLE 3.3 ALUMINA PRECOATED PLATES, GLASS SUPPORT, ANALYTICAL AND PREPARATIVE

Manufacturer	Product	Layer Thickness (μ)	Binder	UV Indicator	Plate Size
Analabs, Inc.	Anasil AG	250	10% CaSO$_4$	–	a, b, c
		500			a
		1000			a
	Anasil AGF	250	10% CaSO$_4$	+	a, b, c
		500			a
		1000			a
Analtech	Alumina G	250	CaSO$_4$	–	a-e, m
		500			a-e
		1000			a-e
	Alumina GF	250	CaSO$_4$	+	a-e, m
		500			a-e
		1000			a-e
	Alumina H	250	–	–	a, b, c
	Alumina HF	250	–	+	a, b, c
	Woelm alumina neutral	250	–	–	a, b, c
	Woelm alumina basic	250	–	–	a, b, c
	Woelm alumina acid	250	–	–	a, b, c
Brinkmann	Aluminum oxide	250	+	–	a, c
	Aluminum oxide/ UV-254	250	+	+	a, c
		1000			a

Supplier	Product				
	Aluminum oxide F-254, EM, Type E	250	+	+	a, c
	Aluminum oxide F-254, EM, Type T	250 / 1500	+	+	a, c
Camag, Inc.	Aluminum oxide D-B	250	+	−	a
	Aluminum oxide DF-B	250	+	+	a
E M Laboratories	Aluminum oxide 60 F-254	250	organic	+	a, c
	Aluminum oxide 150 F-254	250 / 1500	+ / +	+ / +	a, c / a
New England Nuclear	Alumina AG	250	+	−	a, c
	Alumina AGF	250	+	+	a, c
Quantum	Aluminum oxide Q3	250 / 500 / 1000	inorganic	−	a, b, c / a / a
	Aluminum oxide PQ3				
	Aluminum oxide Q3F	250 / 500 / 1000		+	a, b, c / a / a
	Aluminum oxide PQ3F				

TABLE 3.4 CELLULOSE PRECOATED PLATES, GLASS SUPPORT, ANALYTICAL AND PREPARATIVE

Manufacturer	Product	Layer Thickness (μ)	UV Indicator	Plate Size
Analabs, Inc.	Anasil C (Avicel)	250	-	a, b, c
		500		a
		1000		a
	Anasil CF	250	+	a, b, c
		500		a
		1000		a
	Anasil PEI	250	-	a, c
	Anasil MN 300	250	-	a, b, c
	Anasil MN 300F	250	+	a, b, c
Analtech, Inc.	Avicel	250	-	a-e, m
		500		a-e
		1000		a, b, d, e
	Avicel F	250	+	a-e, m
		500		a-e
		1000		a, b, d, e
	MN300	250	-	a-e, m
		500		a-e
	MN300F	250	+	a-e, m
		500		a-e
	Cellulose DEAE	250	-	a, b, c
	Cellulose Ecteola	250	-	a, b, c
	Cellulose CM	250	-	a, b, c
	Cellulose P	250	-	a, b, c
	PEI Avicel	250	-	a, b, c

	10% Acetylated cellulose	250	–	a, b, c
	20% Acetylated cellulose	250	–	a, b, c
Applied Science Laboratories	Microcrystalline cellulose	250	–	a, c
Brinkmann	Cellulose MN300	100		a
		250		a
		500		a
		1000	–	a
	Cellulose MN300/UV254	100	+	a
		250		a
		500		a
	Cellulose Avicel	250	–	a–d
	Cellulose F, Avicel	250	+	a, b, c
	PEI cellulose	250	–	a
Camag, Inc.	Camag cellulose, microcrystalline, DS-0	250	–	a
	Camag cellulose, microcrystalline, DSF-0	250	+	a
	PEI cellulose microcrystalline	100	–	a
	PEI cellulose microcrystalline F	100	+	a
EM Laboratories	Cellulose	100	–	a
	Cellulose F-254	100	+	a
	PEI cellulose F	100	+	a
New England Nuclear	Cellulose C (Avicel*)	250	–	a, c
	Cellulose CF (Avicel*)	250	+	a, c

TABLE 3.4 (continued)

Quantum	Microcrystalline cellulose Q2 (Avicel*)	250 / 500 / 1000	–	a-d / a / a
	Microcrystalline cellulose Q2F (Avicel*)	250 / 500 / 1000	+	a-d / a / a
Supelco	Redi-Coat C, cellulose with SiO_2 binder for amino acid separations	500	–	a
Schleicher and Schuell	Cellulose	100	–	a
	Cellulose/LS 254	100	+	a
	Cellulose, acetylated 20-25%, starch binder, 4%	100	–	a
	Cellulose, acetylated 40-45%, starch binder, 4%	100		a
	Cellulose with Dowex 2-X-8, 5% Dowex	100	–	a
	Cellulose with Dowex 2-X8, 10% Dowex	100		a
	Cellulose with Dowex 50W-X8, 5% Dowex	100	–	a
	Cellulose with Dowex 50W-X8, 10% Dowex	100		a
	PEI cellulose	100	–	a
	PEI cellulose/LS 254	100	+	a

* Avicel is a trademark of the FMC Corporation for their brand of microcrystalline cellulose.

TABLE 3.5 KIESELGUHR PRECOATED PLATES, GLASS SUPPORT

Manufacturer	Product	Layer Thickness (μ)	UV Indicator	Plate Size
Analtech, Inc.	Kieselguhr G (CaSO$_4$ binder)	250	-	a, b, c
Brinkmann	Kieselguhr F-254	250	+	a
EM Laboratories	Kieselguhr F-254	250	+	a
Quantum	Kieselguhr	250	-	a

TABLE 3.6 POLYAMIDE PRECOATED PLATES*, GLASS SUPPORT

Manufacturer	Product	Layer Thickness (μ)	UV Indicator	Plate Size
Schleicher and Schuell	Polyamide with 4% starch binder	120	-	a
	Polyamide with indicator	120	+	a

* Concentrated acid cannot be used as an indicator with these plates.

TABLE 3.7 SILICA GEL PRECOATED PLATES, PLASTIC SUPPORT

Manufacturer	Product	Layer Thickness (μ)	Binder	UV Indicator	Plate Size
J. T. Baker*	Silica gel IB	200	+	-	a, c, e, m
	Silica gel IB-F	200	+	+	a, c, e, m
	Silica gel IB2	200	+	-	a, c
	Silica gel IB2-F	200	+	+	a, c
Brinkmann	Silica gel G	200	+	-	a, c, e
	Silica gel G/UV-254	200	+	+	a, c, e
	Silica gel G, hydro-phobic	200	+	-	a
	Silica gel N-HR (MN)	200	+	-	a, c, e
	Silica gel N-HR/UV-254	200	+	+	a, c
Camag, Inc.	S&S silica gel	250	+	-	a, e
	S&S silica gel F	250	+	+	a, e .
Eastman	Chromagram silica gel	200	polyvinyl alcohol	-	a
	Silica gel with indicator	200	polyvinyl alcohol	+	a
EM Labo-ratories	Silica gel 60	250	+	-	a, c
	Silica gel F-254	250	+	+	a, c
	Silica gel	200	+	+	20×500 roll

| Schleicher and Schuell (S&S) | Silica gel | 200 | organic | – | a |
| | Silica gel with indicator | 200 | organic | + | a |

* All J. T. Baker products are called Baker-flex.

TABLE 3.8 ALUMINA PRECOATED PLATES

Manufacturer	Product	Layer Thickness (μ)	Binder	Indicator	Plate Size
J. T. Baker*	Aluminum oxide IB	200	+ inert	–	a, c, e, m
	Aluminum oxide IB-F	200	+ inert	+	a, c, e, m
Brinkmann	Aluminum oxide N, MN brand	200	+	–	a, c, e
	Aluminum oxide N, MN brand, UV-254	200	+	+	a, c, e
Eastman	Alumina with indicator	200	polyvinyl alcohol	+	a
EM Laboratories	Aluminum oxide 60 F-254	200	+	+	a

* All J. T. Baker products are called Baker-flex.

TABLE 3.9 CELLULOSE PRECOATED PLATES, PLASTIC SUPPORT, ALL BINDER-FREE

Manufacturer	Product	Layer Thickness (μ)	UV Indicator	Plate Size
J. T. Baker*	Cellulose	100	–	a, c, e, m
	Cellulose F	100	+	a, c, e, m
	Cellulose, microcrystalline	100	–	a, c
	Cellulose F, microcrystalline	100	+	a, c
	Cellulose AC10 (10% acetylated)	100	+	a, c
	Cellulose CM	100	–	a, c
	Cellulose DEAE	100	–	a, c
	Cellulose Ecteola	100	–	a, c
	Cellulose PEI	100	–	a, c
	Cellulose PEI-F	100	–	a, c
Brinkmann†	Cellulose MN300	100	–	a, c, e
	Cellulose MN300/UV-254	100	+	a, c, e
	Cellulose MN400, Avicel microcrystalline	100	–	a, c, e
	Cellulose MN 400/UV-254	100	+	a, c, e
	10% acetylated cellulose, MN300	100	–	a
	30% acetylated cellulose, MN300	100	–	a
	CM cellulose, MN300	100	–	a
	DEAE cellulose, MN300	100	–	a, e
	Ecteola cellulose, MN300	100	–	a
	PEI cellulose, MN300	100	–	a,e
	PEI/UV-254	100	+	a

Manufacturer	Product			
Camag, Inc.	S&S cellulose, microcrystalline	100		a
		250	–	a
	S&S cellulose, microcrystal-line F	100	+	a
		250		a
	PEI cellulose, microcrystalline	100	–	a
	PEI cellulose, microcrystal-line F	100	+	a
Eastman	Cellulose	200	–	a
	Cellulose, fluorescent indicator	200	+	a
EM Laboratories	Cellulose	100	–	20×500 roll
		100	–	a
	Cellulose F-254	100	+	a
	PEI cellulose F	100	+	a
Schleicher and Schuell	Cellulose, microcrystalline	100	–	a
	Cellulose, microcrystalline with indicator	100	+	a
	PEI cellulose	100	–	a
	PEI cellulose with indicator	100	+	a

* All products called Baker-flex.

† All plates coated with MN brand cellulose.

71

TABLE 3.10 POLYAMIDE PRECOATED PLATES, PLASTIC SUPPORT, NO BINDER

Manufacturer	Product	Layer Thickness (μ)	UV Indicator	Plate Size
J. T. Baker*	Polyamide 6	100	–	a, c
	Polyamide 6F	100	+	a, c
	Polyamide 11	100	–	a, c
	Polyamide 11-F	100	+	a, c
Brinkmann	Polyamide 6	100	–	a
	Polyamide 6/UV-254	100	+	a
Camag, Inc.	Polyamide and acetylated cellulose F	130	+	a
	Micropolyamide, double-faced	25 both sides	–	5×5, 15×15 cm
Schleicher and Schuell	Micropolyamide, double-faced	25 both sides	–	a

* All J. T. Baker products are called Baker-flex.

TABLE 3.11 ION-EXCHANGE PRECOATED PLATES*, PLASTIC SUPPORT

Manufacturer	Product	Layer Thickness (μ)	UV Indicator	Plate Size
Brinkmann	Ionex-25 SA-Na, strong acid, Na$^+$ active	250	-	a
	Ionex-25 SB-Ac, strong basic, CH$_3$COO$^-$ active	250	-	a
	Ionex-25 SB-Ac/UV 254, strong basic, CH$_3$COO$^-$ active	250	+	a

* All are manufactured by Macherey, Nagel and Co. and distributed by Brinkmann.

TABLE 3.12 SILICA GEL PRECOATED PLATES, ALUMINUM SUPPORT

Manufacturer	Product	Layer Thickness (μ)	Binder	UV Indicator	Plate Size
EM Laboratories (E. Merck plates)	Silica gel 60	250	+	-	a, c
	Silica gel 60 F-254	250	+	+	a, c
	Silica gel 60/ kieselguhr F-254	250	+	+	a
ICN Pharmaceuticals (Woelm plates)	Silica gel	200	organic	-	a
	Silica gel F-254/366	200	organic	+	a

TABLE 3.13 ALUMINA PRECOATED PLATES, ALUMINUM SUPPORT

Manufacturer	Product	Layer Thickness (μ)	Binder	UV Indicator	Plate Size
EM Laboratories	Aluminum oxide 60 F-254	200	+	+	a
	Aluminum oxide 150 F-254	200	+	+	a

TABLE 3.14 CELLULOSE PRECOATED PLATES, ALUMINUM SUPPORT

Manufacturer	Product	Layer Thickness (μ)	UV Indicator	Plate Size
EM Laboratories	Cellulose	100	-	a, roll
	Cellulose F-254	100	+	a
	PEI cellulose F	100	+	a

TABLE 3.15 KIESELGUHR PRECOATED PLATES, ALUMINUM SUPPORT

Manufacturer	Product	Layer Thickness (μ)	UV Indicator	Plate Size
EM Laboratories	Kieselguhr F-254	250	+	A

TABLE 3.16 SILICA GEL PLATES, IMPREGNATED GLASS FIBER

Manufacturer	Product*	UN Indicator	Plate Size
Gelman	Silica gel SG ITLC-SG	-	a, c
	Silicic acid SA ITLC-SA	-	a, c
	Silicic acid SAF ITLC-SAF	+ (254 and 350 nm)	a, c
	SAF-D drug identification media	+	a

* ITLC is a trademark of the Gelman Co. and is an abbreviation for instant thin layer chromatography.

CHAPTER 4

Preparation and Application
of the Sample

4.1 INTRODUCTION

The field of TLC has increased considerably in recent years. However, in spite of the numerous reports on methodology, there has
been no real treatise on the art of preparing a sample and applying it to a thin layer. The technique appears simple enough, so
much so that one can tend to become careless and then wonder why
the results are not satisfactory. TLC can generally be done with
less preparation and attention than any other separation method.
But, for reproducibility and for quantitative determinations as
well as mass screening, some attention to detail is required.

The room used for TLC should be of constant temperature and
humidity. It must be clean and free of dust, particularly in industrial areas, and well ventilated; open tanks release many solvent fumes. Dust particles interfere with fluorometric evaluation
of TLC because many dust particles are fluorescent, some react
with the reagents used for visualization, and lint particles from
clothing are often fluorescent. Since TLC is essentially a micro-
technique, extreme cleanliness should be in order, particularly
when quantitation is to be performed.

The environment should also be free of chemical fumes that
may alter the sample or be adsorbed into the plate while the sample is being applied.

With the sample in hand and a micropipet of some sort and the
proper plate before you, what more simple technique for separation
is there anywhere? When you dip the pipet into the sample, it is
easy to see the microcapillaries fill. Microsyringes need some
manipulation to be sure they are filled. Touch the sides of the

pipet to the container to remove excess solution.
Now carefully lower the tip of the delivery device to the
layer. Capillary action removes the solution from the glass capillary. Plunger action drives it from syringes. The amount of
solution to be transferred to the layer determines the procedure
from here. It is better to apply small proportionate amounts repeatedly. The aim is to form the smallest spot amenable with the
solvent.
Now that the plate has a spot or streak or a series of them,
it can be put in the tank.
All appears very simple. However, there is more to it than
just this. This chapter examines aids to spotting, pitfalls of
sample application, the different types of delivery systems (both
manual and automatic), selection of solvents for application, and
the reasons certain spots appear the way they do on finished chromatograms.

4.2   SOLVENT SELECTION

The first consideration before any chromatography can be done is
to decide which solvent should be used to dissolve the sample.
Often the analyst receives a sample whose physical characteristics and solubility are not known. If the general nature of the
compound is known, experience will tell which solvent to use as
a starting parameter. Reference books such as *The Merck Index*
and *The Handbook of Chemistry and Physics* are extremely useful in
this regard and should be in the library of every chromatographer.
However, when the sample is a derivative, the solubility characteristics can be drastically changed, or there may not be enough
sample to perform the general solubility tests. If the material
is crystalline, its solubility can be determined readily with the
aid of a microscope. Place a few crystals of the sample on a microscope slide and mount the slide in the field of the microscope.
While viewing the crystals, use a micropipet to place a drop of
solvent on the crystals. If the crystals do not dissolve, evaporate the solvent and select another for testing. Judicious testing of a number of solvents will expedite selection of the optimal
one for the experimental analysis.
The process of application of the sample to the TLC may be
one of the most important steps for the success of the separation.
Proper preparation of the sample itself is prerequisite to application of the sample to the layer. The nature of the sample,
or the components in the sample that is to be analyzed, determines
the actual extraction or preparation of the compound in question.
This chapter will cover the handling of the sample before and dur-

ing its application to the chromatogram. Extraction of the compounds of interest from a complex sample, and dissolution after-
wards in proper vehicles (solvents) depend on the compound class
under consideration. Polar compounds do not dissolve well in nonpolar solvents. Nonpolar compounds will not dissolve in polar
solvent. Usually ionic compounds dissolve best in a polar solvent
to which the opposing ion has been added. Generally speaking,
however, for good application of samples to the layers the solvent
used should be highly volatile and as nonpolar as possible. The
reason for this is that nonvolatile solvents spread through the
plate layer matrix during their period of contact, taking the sample with them.

Likewise, a polar solvent will spread its solute through the
matrix because of high solvent power as well as low volatility.
Samples can, however, be changed to derivatives prior to chromatography to facilitate solution in solvents of lower polarity and
higher volatility. This also can aid in detection, identification,
or analysis. Many polar compounds are difficult to chromatograph
because of their poor solubility and high polarity. Cases have
been known where a nonpolar solvent had been used to handle a polar solute and the operator wound up wondering why no spots appeared on the plate in spite of following directions described
in the literature. Basically, a knowledge of the compound to be
separated aids in selection of the sorbent, the developing solvent
(mobile phase), and the solvent to be used for solutions and during application of the sample.

The terms polar and nonpolar deserve comment. A polar solvent is one of high ion potential. Conversely, a nonpolar solvent
is one with very low ion potential. Water and acids or bases are
highly polar, whereas hydrocarbons are solvents of the nonpolar
class. The eluotropic series described in detail in Chapter 5
lists solvents in order of polarity as well as power of elution
from a sorbent. This type of table is sometimes useful in selecting a solvent for the most favorable sample application conditions.

4.3 SAMPLE APPLICATION

Samples should be applied as a 0.01-1.00% solution in the least
polar solvent in which they are soluble. Normally, 1-5 µg/µl solution concentrations are practical. Solvents with too high a
boiling point or high polarity are difficult to remove from the
sorbent during application. If a small amount of solvent is retained in the sorbent after application of the sample, it will ad-
versely affect the separation, due to spreading of the sample in
the matrix. Be sure the samples are dry before placing the plate

in the developing chamber. Some knowledge of the relative concentration of the components under examination is usually helpful in
this respect. This will determine the smallest amount of sample
or extract that can be applied to the plate. Dilute solutions can
be applied to the layer with solvent drying between successive applications, or they can be concentrated. Overspotting is a useful
technique, but care must be exercised since too rapid an application of sample results in "rings"; that is, the solvent washes the
sample to the outer edges of the application area. This sometimes
results in spurious separation of two zones where there should be
only a single spot, particularly if the ring formed at the starting line has a wide diameter. Sometimes drying between applications with a hair dryer or a stream of filtered air will facilitate sample application, but only if the sample components to be
separated are nonvolatile; otherwise they will be lost. The zone
of application should be as narrow as possible, or the spots
should be as small as is compatible with the concentration. Spots
that are too highly concentrated will give poor separation, since
the mobile phase solvent tends to flow through the point of least
resistance and will travel around the spot. This will result,
after development, in separated spots that are unsymmetrical.
Ideally, if the spots are properly applied the separated zones are
symmetrical and compact. The spots tend to diffuse and spread
somewhat as they migrate further up the plate (at increasing R_f).
Width of separated zones is a function of retention in the system.
In gas chromatography as well as in liquid chromatography, the
substances of longer retention time show wider baselines in the
recorded peaks. Recordings of scans of TLC show larger base
widths from the substances with high R_f. Applying the sample as a
narrow streak has an advantage over spot application; a streak
usually produces a sharper separation, which in turn gives improved reproducibility. The mobile phase is less likely to flow
around a streak than a spot during development.

Samples should be applied near one side of the layer, usually
about 1.5-2 cm from the edge, so that the layer makes contact with
the mobile phase but the sample zone itself is not immersed in the
liquid. This avoids dissolution of the sample in the developing
phase. In practice a sheet of white cardboard can be used as a
background while the sample is applied to the chromatoplate. A
dark line can be drawn on this background so that when the plate
is positioned on it the line will be seen through the layer 2 cm
(or the predetermined distance) from and parallel to the edge.
This serves to provide a uniform marker across the layer to guide
the sample application.

The sample may be applied as a spot or as a series of spots
to form a streak. The application of a continuous streak is often
used, made possible with modern instrumentation. Streaking of the

sample along the baseline is not easily performed manually, but it has been made easier by the use of the trough device of Achaval and Effelson (1) or the automatic applicator described by Coleman (2). Automatic devices for streaking, however, overcome this and are described later in this chapter.

The band width or spot diameter must be as small as possible, since as the chromatogram develops, the areas occupied by the separated zones tend to increase and the band or spot widens due to diffusion. Nyborn (3) has observed that there is no direct relationship between the amount per spot and the spot size variations being due to characteristics of the separated substances and to layer thickness within limits. The load applied is critical and affects the shape and position of the zone after development. Overloaded sample applications produce cometlike vertical streaks. If the adsorption isotherm is convex, the R_f measured from the midpoint of the streak will be greater than normal. If the adsorption isotherm is concave, the R_f value will be less than normal. The ends of the streak will overlap other spots on the chromatogram, rendering isolation and resolution poor. This is particularly true when one compound is present in larger quantity than another with a similar R_f value. This is illustrated in Figure 4.1. Herein lies one of the most common faults of the practice of thin layer chromatography. It should be remembered that this is a sensitive technique, and the smallest sample size that can show visualization of all desired components is the one to be applied. This is one of the prerequisites for reproducible R_f values. This general rule also holds true for sample application in quantitative work. Too concentrated a zone will tend to give nonlinear calibration curves.

The technique of applying microliter amounts of solutions is not easy. The accuracy of methods of delivering small volumes has already been discussed. In practice, depending on the application instrument used, small drops do not leave the tip of a needle in contact with a thin layer, and care must be used to avoid disturbing the layer. A hole in the layer may cause distorted triangular or crescent spots in the final chromatogram (see Figure 4.2). Truter (4) describes this type of sampling effect.

Techniques for applying samples to TLC plates under specialized conditions such as from a gas chromatograph are discussed in Chapter 14.

4.4 MANUAL SAMPLE APPLICATORS

It is unfortunate that in many reports of quantitative TLC methodology the preparations of the solutions, the solvent used, and the pipets used for applying the sample itself are not described in

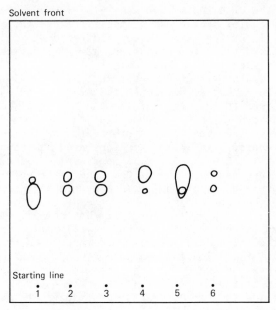

Load (μg)

	1	2	3	4	5	6
Upper spot	0.5	2	3	5	15	0.5
Lower spot	10	4	3	1	0.5	0.5

Solvent system: isooctane:acetone:n-butanol
(58:33.6:8.4); silica gel G layers.

Figure 4.1 The effect of load on the resolution of a mixture of
reserpine and rescinnamine.

detail. Some expression of this methodology sometimes is necessary for the success of the separations as well as the qualitative and quantitative results.

     For much general work where exact sample size is not the con-
sideration, capillaries fabricated in the laboratory are often
used. These are readily made as they are needed. However, they
are not suitable for quantitative work or when duplication of the
size of the zones on the starting line is desired. There are

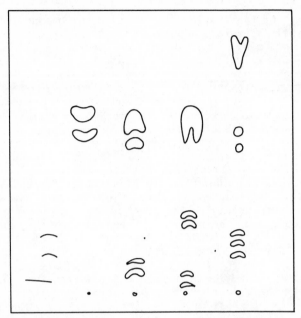

Figure 4.2 Distorted shapes of spots, due to excessive pricking of the surface of the adsorbent.

many types of capillaries in a variety of forms. It is even possible, if a concentrated sample is available, to transfer it as a
drop from a glass rod. However, these miscellaneous means of
transfer should be considered only for those who do not have to
reproducibly apply samples.
     The sample can also be applied with micropipets or melting
point capillaries. There are a number of automatic spotters of
varying design available commercially. Calibrated disposable
capillaries as well as ordinary capillaries are generally available. These are preferable for most routine work when few samples are involved, the advantages being the saving of time as well
as avoidance of sample contamination. Since chromatographic methods used today utilize only small test samples, 1 μl or even 0.1
μl sample solutions will suffice for a complete qualitative or
quantitative analysis. The analyst is faced with the problem of
delivery precision of the various micropipets. This poses a
source of error that may be greater than the combined errors in
the chromatographic process itself. The repetitive spotting of
microliter volumes is a tedious operation, particularly if a
large number of samples must be analyzed. Automatic spotters have
become increasingly useful in this regard.

In spite of the availability of a number of sample delivery systems, little attention seems to have been given to reproducibility or determination of whether the measuring instruments actually do deliver replicates of constant volume. Volumes of 1-5 µl are frequently used, and therefore errors in repetitive delivery would invalidate any attempts to reproduce identical conditions from day to day. A certain amount of experience is necessary to obtain maximum accuracy. Fairbairn and Ralph (5) report details of the results of repetitive measurement of volumes by a number of workers, each using a favorite syringe or micropipet. These results are given in Table 4.1. The participants used radioactive solutes to facilitate the determination. Each delivered 10-16 replicates (a) directly into glass vials for counting or (b) onto thin layer plates or paper chromatograms. Areas of the paper or sorbent containing the zone (from b) were transferred to vials for counting. The results as seen in the table indicate that pipetting errors ranged from ± 6 to ±20% for individual measurements. This is indeed a large source of error in quantitative work as well as in repetition of sample analysis.

Emanuel (6) more recently reported results of a survey of reproducibility in delivery by a number of micropipets. It appears from this report that simple glass capillary pipets may be the most accurate microdelivery system. Since they are disposable and relatively inexpensive, they are useful in applying samples to thin layer plates. Table 4.2 shows the results of this study. These experiments evaluated reproducibility of delivery by the pipets, not the ease of delivering the sample to the layer.

A number of companies produce capillary pipets for general use. Disposable uncalibrated glass capillaries are widely used for applying samples to layers. The "Microcaps" produced by Drummond Scientific Co. can be obtained both calibrated and uncalibrated. These pipets are simple to use. They fill readily, and once touched to the layer at a *right angle*, the capillary action of the layer draws out the liquid completely. Since they are disposable, the time-saving factor is considerable and contamination between samples is eliminated.

An interesting apparatus using glass capillaries was described by Flinn (7). The apparatus was essentially designed for use with samples from tablets or capsules of pharmaceutical formulations. It is a technique that provides maximum avoidance of sample contamination or decomposition. The apparatus will handle 28 samples simultaneously in a single operation. The powdered sample is loaded into an end of a 10 µl micropipet by plunging it into the powdered sample material. The operation is repeated as often as necessary to load sufficient sample in the pipet. The same end of the pipet is then repeatedly plunged into a 2 mm layer of sorbent. Three parts of sorbent to one part of sample was

TABLE 4.1 ERRORS DUE TO MEASUREMENT OF SMALL VOLUMES OF SOLUTION BY EXPERIENCED
WORKERS, EXPRESSED AS COEFFICIENT OF VARIATION
(S.D. OF INDIVIDUAL RESULTS CALCULATED AS A PERCENTAGE OF THE MEAN)

| | | Coefficient of variation | |
Worker	Solute	Volume of solution measured (μl)	Direct into phosphor (%)	Via adsorbent (%)
A	Morphine-2-T	10 (Agla)	7.8	7.3 (paper)
B	Morphine-2-T	10 (Agla)	5.8	10.1 (paper)
C	Morphine-2-T	2 (micropipet)	6.6	3.3 (paper)
C	d-Glucose-^{14}C(U)	2 (micropipet)	4.7	
		5 (micropipet)	2.5	
D	d-Glucose-^{14}C(U)	1 (Hamilton)	11.1	8.2 (paper)
			10.4	7.8 (paper)
		1 (Hamilton)		8.8 (TLC)
				9.8 (TLC)
B	d-Glucose-^{14}C(U)	5 (Agla)	5.5	3.2 (paper)
E	d-Glucose-^{14}C(U)	5 (Agla)	3.4	6.0 (TLC)
F	l-Tyrosine-^{14}C(U)	5 (Agla)	4.5	9.1 (paper)
		10 (Agla)	10.9	11.9 (paper)
B	l-Tyrosine-^{14}C(U)	5 (Agla)	7.9	
G	l-Tyrosine-^{14}C(U)	5 (Agla)	4.7	

Table 4.2 ERRORS DUE TO MEASUREMENT OF SMALL VOLUMES OF SOLUTION
USING THE MACHINE REFERRED TO IN THE TEXT

Worker	Solute	Volume of solution measured (μl)	Coefficient of variation	
			Direct into phosphor (%)	Via adsorbent (%)
B	Morphine-2-T	2	2.4	1.8 (paper)
B	Morphine-2-T	5	2.2	2.5 (paper)
B	Morphine-2-T	9.6	1.6	2.1 (paper)
B	d-Glucose-^{14}C(U)	5	2.0	2.7 (paper)
B	l-Tyrosine-^{14}C(U)	5	1.5	1.5 (paper)

found to be enough to filter the sample. The pipets are then positioned in the holder lying on its side. The holder is then placed upside down over a tank of solvent; all the micropipets are filled with the solvent simultaneously by placing the empty ends in the solvent long enough for them to fill by capillary action. Then the holder is placed over the TLC plate, and the loaded pipets are pressed against the plate so that the solvent flows down through the sample, thus extracting the active ingredient and depositing it on the layer. The plate is then developed in the desired manner. This may not necessarily be a quantitative method, but it does treat each of the samples in the same manner and gives comparison under the same relative conditions.

Since most samples for separation are small, the containers that hold the sample as it is being applied to the layer must be discussed. The capillary pipets discussed so far are of limited length, usually 5-10 cm. This would in some cases limit the size of the container used to handle the sample. Very convenient sample holders are glass-stoppered microcentrifuge tubes of 0.5, 3, and 5 ml volume. Most solvents used for transfer of the sample (by definition) should be highly volatile for ideal spot size. This means that in quantitative TLC it is necessary to have on hand containers that are easy to open and close to prevent undue evaporation of the solvent while samples are being applied.

For easy handling of the glass capillary during sample transfer, a holder can easily be made, or those commercially available should be used. A set of capillaries supplied by Drummond Scientific is accompanied by one of these holders as pictured in Figure 4.3. A piece of 8 mm glass tubing can be cut to the length desired, and the medicine dropper bulb with a small hold in the top is placed on one end. A small rubber stopper, cut to fit the base at the other end (such as from a Microcap holder), is pierced to form a hole through which the capillary pipet fits tightly. The apparatus so formed facilitates handling of the sample and is useful in withdrawing sample from larger test tubes. The bulb facilitates sample uptake when the capillaries are long. The device is particularly useful in forcing out the total sample as well as in controlling the rate of dispersal.

The capillary pipets available from Drummond Scientific can also be used in a transparent barrel graduated in 0.1 μl divisions. The disposable bore is the precision glass capillary, which takes a stainless steel plunger. The plunger tip travels the length of the bore and is used as the reference point for readings. The bores (capillaries) and plunger are inexpensive enough to justify discarding after a single use to avoid contamination of samples and the time-consuming cleanup. The finger loop handle of the plungers allows one-hand operation (Figure 4.4).
The capillary passes through a Teflon collar at both ends of

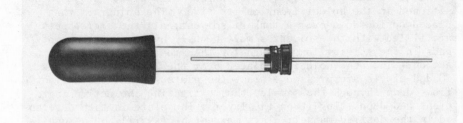

Figure 4.3. Glass capillary micropipet in holder. Courtesy A. H.
Thomas Co.

the plastic barrel and is held by a collet chuck. The wire plunger is held in place by a springlock clamp of the handle. These
micropipets are supplied by Drummond under the name Microtol, and
are available in sizes of 15 µl and 20 µl with graduations of 0.1
µl. If larger sizes are required, the syringe type adaptation is

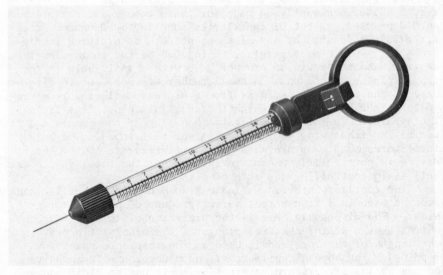

Figure 4.4 Disposable glass barrel microsyringe. Courtesy A. H.
Thomas Co.

available. The reusable stainless steel plunger has a Teflon tip
and finger loop for one-hand operation. The disposable bores are
precision glass capillaries held in place in a glass barrel by
threaded plastic end fittings. The glass barrel is graduated in
1 μl divisions for the 100 μl capillaries.

Glass capillaries are stocked by most chemical supply firms.
Since they are inexpensive, they are the pipets of choice when
large numbers of samples are to be applied, particularly in qualitative work. They are widely used in drug screening work where
TLC is the most commonly used analytical procedure.

Disposable capillaries are ideal when radioactivity is involved. Their use avoids cross contamination, the bane of those
engaged in metabolic studies. The conventional use of labeled
internal standards as well as reference substances added for assessment of recovery is facilitated when these capillaries are
used for sample transfer. As pointed out earlier, they are among
the most accurate transfer vehicles for use in TLC.

Microsyringes such as those manufactured by Hamilton Co.,
Unimetrics, and Labcrest, among others, have seen wide acceptance
and can be used for delivery of calibrated volumes of solution.
They usually are made with glass barrels and a stainless steel or
tungsten plunger. Some have a replaceable needle; others have
permanently attached needles. They are available in volumes of
1 μl to 100 μl and are calibrated to deliver as little as 0.05 μl.
They were originally designed for use in gas chromatography and
are equally adaptable to use in TLC. These syringes require some
practice in use for reproducibility of sample delivery in quantitative work. Figure 4.5 shows an example of this type of syringe. With care and proper preparation of standard solutions it
is possible to obtain reproducibility in the separations with a
precision better than ±1% with the same syringe. These pipets
must be meticulously cleaned between the application of repetitive samples.

Figure 4.5. Microsyringe with a capacity of 1 μl. Courtesy A. H.
Thomas Co.

   Smooth delivery of the samples is hindered by bores and
plungers that are not clean. Spurious zones on chromatograms can
be due to solutes carried over from sample to sample in improperly
cleaned pipets. An uneven and sporadic travel of the plunger
through the bore prevents smooth application of the sample. With
extraneous dirt buildup in the bore, the plunger sometimes becomes
frozen.
   Microsyringes are readily cleaned by drawing up fresh solvent
with the plunger and dispelling it, repeating this a number of
times. The preferred method is to remove the plunger and wipe it
with a lint-free tissue saturated with solvent. Then, using a
water aspirator with a rubber hose, draw a volume of clean solvent
through the needle and barrel. This method is used with nondis-
posable glass micropipets.

   4.5 AUTOMATIC SAMPLE APPLICATORS

Any treatise on sample application would not be complete without
a discussion of automatic sample applicators. There are a number
of these commercially available. These instruments can be divided into four types:

   1. Holders for x number of disposable capillaries.
   2. Syringe-type sample streak applicators.
   3. Large capacity tube fitted with a capillary delivery system to deliver a single sample.
   4. Multiple syringe holders with provision for automatic de-
pression of the plunger.

   A holder with 19 individual capillaries has already been described. This device is now available commercially (A. H. Thomas
Co.) and is shown in Figure 4.6. A description from their catalog follows:
   "U-shaped base A of applicator accommodates sample trough
B or micro tube rack C (and subsequently the coated plate G)
beneath perforated bar for self-filling capillary spotting
pipets D. Bar H is supported by coil springs on guide rods at
each end. Bar H is pressed down to lower the pipets for fill-
ing and for delivering samples on page 20 mm from edge.
fit loosely in holds and exert only their own weight on coating.
   "Micro tube rack takes 19 tubes approximately 6 mm o.d.
Suitable tubes can be improvised by sealing one end of short
length of glass tubing. Length 8 inches; holes are 1/2 in ×
3/16 in deep. Nickeled brass.
   "Sample trough is V-shaped, 7-5/8 in long, width at top
3/16 in, depth 1/2 in. Stainless steel."

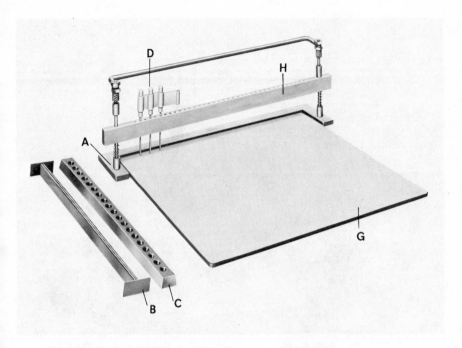

Figure 4.6.   Thomas-Morgan sample spotter.   For description see
text.   Courtesy A. H. Thomas Co.

The individual samples can be added to the perforated bar
and all samples taken up at once, or the pipets may be used to
withdraw samples from the original containers.

An example of a streak applicator is shown in Figure 4.7.
This apparatus can be used for the uniform application of sample
onto TLC plates as a narrow band for quantitative separations and
preparative layer chromatography.   Essentially it is a carriage-
mounted syringe pipet on a horizontal track with an adjustable,
inclined bar device for precise linear delivery of sample in proportion to pipet travel.   Plates are placed directly on the base
of the streaker stand.

The included syringe pipet, 100 μl capacity, has a delivery
needle 65 mm long with square-cut tip.   As the pipet carriage is
advanced manually across the coated plate, a plunger drive-weight
assembly travels downward along an inclined bar, resulting in
sample delivery at a linear rate.   The adjustable angle of incline
governs the delivery rate.   A reference scale is provided for re-
setting angle.

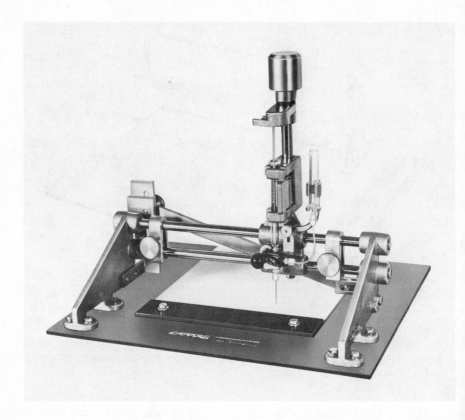

Figure 4.7.   Camag Chromatocharger for streaking samples on TLC plates.   Courtesy Camag Inc.

When a streak length of less than 40 mm is required, a short band accessory must be employed. This device, available on special order, positions masking strips in streak path to produce bands 10 to 40 mm long.

Streakers are not applicable when multiple screening of many samples is required. They are limited to applying relatively large samples as a streak across the origin of the plate. In general they work well, and resolved areas on developed chromatograms consist of well-defined linear zones across the plate.

Figure 4.8 shows streaking device consisting of a syringe with a driver for the plunger synchronized with the motor, which moves the syringe over the starting line on the plate. The extra-long needle is positioned so that the point just delivers the sol-

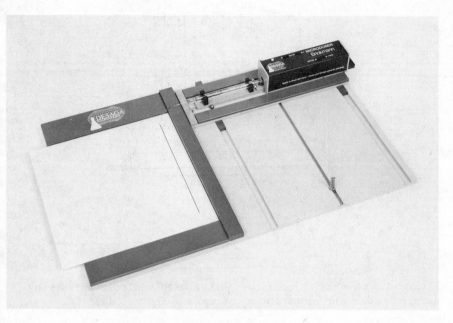

Figure 4.8.   Brinkmann Desaga motor syringe.   Sample streaker for
TLC.   Courtesy Brinkmann Instruments.

vent to the layer.

The sample applicator recently introduced by Kontes Instruments Division, shown in Figure 4.9, essentially consists of
large-bore glass tubes fitted with needles.   The needle is positioned just above the layer to deliver the solute gradually to
the layer by capillary action.   The essential feature is that the
needle is surrounded by a template with orifices that permit compressed gas flow over the layer to evaporate the solvent as the
solution percolates into the layer.   The gas (air, nitrogen, etc.)
can be cooled or heated.   Because of this, regardless of the
volume of solvent, a small spot will result.   This type of sample
applicator can be very useful when dilute samples must be applied
as a small zone to the layer.   Several applications of up to six
samples can be performed simultaneously.

The automatic spotter made by Analytical Instrument Spe-
cialties permits application of up to 19 samples simultaneously.
The amount of the sample delivered is determined by the position
of the plunger.   The syringes are clamped horizontally in a
holder.   A bar is placed across the needles to depress them to a
position in which the tip is just above the layer.   A motorized

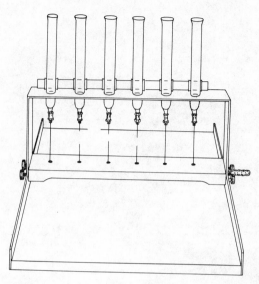

Figure 4.9. Kontes Chromaflex spotter with manifold for drying samples under air or nitrogen. Courtesy Kontes Glass Co.

bar is placed behind the plungers to drive the plunger through the syringe at a predetermined rate. The plungers of the syringes with the larger volumes will be removed the farthest from the bore and will be depressed first by the driving bar. The plungers of the syringes with the least volume to be delivered will be depressed last by the driver. This spotter is equipped with a heater, and the layer can be warmed during sample application. Compressed gas can also be directed over the spots. The whole operation results in reproducible spot size. Temperature, gas flow, and rate of travel of the depressing bar can be easily controlled. Uniform sample application can be obtained from day to day with this type of controlled spotting.

4.6 AIDS AND GUIDES FOR MANUAL SAMPLE APPLICATION

Application of samples to a TLC plate, especially when multiple samples are involved, becomes a tedious operation. Herein may lie one of the difficulties included with reproducible and ideal separation. There are a number of aids to spotting that will help overcome some of these difficulties. The simplest of these is the spotting template available from most suppliers of TLC equipment. The plate is positioned in the apparatus so that one edge is matched below a notched guide along a line (usually 1 cm from the edge) designated as the starting line. There are notches evenly spaced along this edge of the guide. The samples are applied in the notches to form evenly spaced replicate applications along the starting line. One of these is illustrated in Figure 4.10.

Also available from Brinkmann is a calibrated multipurpose TLC template. Made of acrylic plastic, it can be used as a guide to sample application points, and after chromatogram development it can be used for measurement of the R_f values. It has a sliding hairline for easy reading. It may also be used as a storage tray for a plate.

More recently introduced by Sindco Corporation is a unique apparatus known as a "spot-dry sampler" (Figure 4.11). This in-

Figure 4.10. Plastic spotting template for evenly spacing sample application on a TLC plate. Courtesy Camag, Inc.

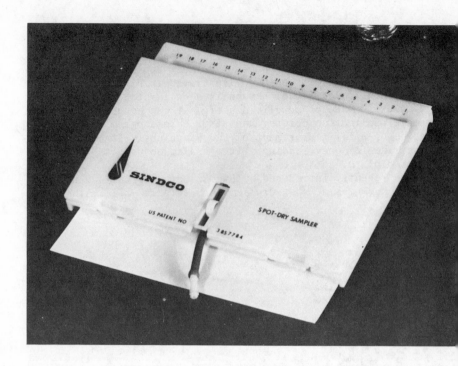

Figure 4.11. Spotting template with gas drying system. Details in text. Courtesy Sindco Corp.

strument is a template with provision for gas flow (cold or warm) to the point of application. The apparatus is designed as a hand rest for chromatographic spotting and features a drying system that allows the gas flow to be easily stopped while applying the sample. When the hand is raised between sample applications, the gas flows over the zone to quickly evaporate the solvent. In this way the application zone (starting spot) is kept as small as possible, a prerequisite for good resolution as well as uniform final zones. This instrument is particularly valuable when application using high boiling solvents is required. It also assures that the application zone is dry before the plate is placed in the development chamber. This apparatus is inexpensive and simple to use.

In some applications there are those who advocate scoring the layer into individual lanes. Plate scorers are available for this purpose, if it is not desired to score the lanes manually using a straight edge and blade, which is difficult to do reproducibly. The one pictured in Figure 4.12 is supplied by Schoeffel Instrument Corp. and divides the 20 cm layer into 20 individual lanes. Applying the samples in the lanes provides a means for uniformly spacing the samples across the starting line. Cross contamination between samples is avoided in close sample application since the samples will not cross the scored lines, which are devoid of sorbent. The scored lanes aid the uniform flow of the mobile phase during development.

4.7   SOURCES OF ERROR IN VOLUME MEASUREMENTS

Fairbairn (8) has discussed at length some of the errors in the use of pipets and syringes in sample application. Among these sources is "creep back" of the solvent as the drop forms on the end of the syringe needle. The following discussion, taken from

Figure 4.12   Layer plate scoring device for producing sample lanes on a TLC plate.   Courtesy Schoeffel Instrument Corp.

his work, should be of aid to the chromatographer.

"The delivery of a number of drops from the needle of an Agla syringe by free fall was observed with a lens and it was noted that from time to time an accumulating drop suddenly slipped up slightly from the point and when the drop finally fell on to the paper a definite proportion remained on the stem. This 'creep back' effect was cumulative and sometimes a sizeable volume remained on the stem for some time; then quite unpredictably it would disappear with a succeeding drop. The creep back effect varied with the solvent used and was particularly noticeable with methanol; occasionally after the delivery of many drops of a methanolic solution of morphine, crystals of the latter substance were seen to have formed a tide mark on the stem of the needle. At times an exceptionally high tide of creep back would reach the crystals, redissolve them, and wash them into a succeeding drop. Accordingly some of the needles used in the experiments recorded in Table 4.1 were washed after each series of radioactive solutions had been delivered. In several cases the 'remainder' on the needle represented a significant proportion of the total radioactive substance measured, e.g. for morphine-2-T (Worker B) after delivering 16 volumes the remainder represented 9.6% of the total delivered. Worker E, on the other hand, avoided creep back by using a hooked needle previously dipped in silicone; there was no detectable remainder after his series of deliveries.

"A second source of error arises from the fact that the measured drop does not always fall freely from the end of the needle as its weight may be insufficient to overcome surface tension effects. To assist this most workers touch the drop on to the surface of the paper or adsorbent or against the glass vial. We found that this process withdrew fluid from the lumen of the needle by capillarity and sometimes it required the delivery of 0.6-0.8 µl of solution from the barrel of the syringe into the needle before the succeeding drop made its appearance. The amount withdrawn varied according to the bore of the needle, time of contact and state of absorbency of the absorbent material, which decreases as more liquid is added to it.

"It is interesting to note that with a micropipet the degree of creep back is greatly reduced since only one delivery is made and capillation is probably taken into account when calibrating. These facts may explain why the errors of Worker C (Table 4.1) were somewhat lower than those of the workers using syringes. As the latter are obviously preferred for replicates or for measuring varying volumes an attempt was made to overcome the two sources of error by (a) siliconing the needles, (b) using different bore thicknesses, and (c) filing the tapered ends till they were at right angles to the long axis, but none of these methods were entirely successful. Success was finally obtained by means of a machine...which automatically delivers small volumes by rapid ejection or throwing. Creep back was prevented and since even quite small drops could be forced on to the paper or adsorbent without touching the surface, capillation was also prevented. This twofold advantage was demonstrated by producing a series of drops from the machine by forcible ejection in the normal way; the coefficient of variation was found to be $\pm2.5\%$. A second series was produced from the machine in identical circumstances except that the needle was brought sufficiently near to the paper for the drops to be drawn off by capillation, that is, to fall rather than be thrown. In these conditions the coefficient of variation rose to $\pm7\%$. Creep back had also occurred as the remainder left on the needle after this series represented 3.27% of the total quantity delivered. When the normal throwing procedure was used the remainder was only 0.32%.

"In Table 4.2 some results obtained with the new machine are shown and it is obvious that the errors are considerably less than those for hand delivered drops (Table 4.1). The results are also more consistent, tedious hand spotting with likelihood of fatigue is avoided and the resulting spots are small and of constant area since equal volumes (0.2-0.4 µl) are ejected each time and the circle of solution allowed to dry before the next drop is ejected."

During the operation of syringes for the delivery of samples, some experience by the operator must be acquired before reproducible solvent delivery will be attained. In context with the two previous examples of error, the contact of the needle tip must be the same each time. The edges should be clean at all times. Possibly more important is the uptake of the solvent into the barrel of the syringe, since it is not possible to see the liquid in the steel needles usually found on syringes. The operator must be sure that there are no air bubbles in the needle. One must be sure that the same type of delivery is used each time. Some touch the tip to the layer during delivery of the sample while pushing down the syringe plunger. Some collect a drop of the solvent on the tip and touch this to the layer. Others hold the tip just above the layer, and the capillary action of the layer draws the liquid while the plunger is driven down. In any of these operations it is necessary to repeat the particular method while avoiding disturbance of the layer. Odd shaping of final spots results when holes or scratches are made in the layer during these operations.

The syringes or delivery devices must be kept clean. Oil film or build up of solute in the bore or on the outer surface will affect solvent delivery. In this respect disposable pipets

have an advantage. The washing of the pipets used in automatic
delivery systems must be through and done in the same manner each
time. Multiple rinsing between deliveries is imperative. The
syringe must be dried between applications, especially in systems
which are all metal, because it is not possible to see the amount
of washing solvent remaining in the needle.

REFERENCES

1. A. Achaval and R. D. Effelson, J. Lipid Res., *7*, 329 (1966).
2. M. H. Coleman, Lab. Pract., *13*, 1200 (1964).
3. N. Nyborn, J. Chromatogr., *28*, 447 (1967).
4. E. V. Truter, *Thin Film Chromatography*, Cleaver-Hume Press
    Ltd., London, 1963.
5. J. W. Fairbairn and S. J. Ralph, J. Chromatogr., *33*, 494
    (1968).
6. C. F. Emanuel, Anal. Chem., *45*, 1568 (1973).
7. P. E. Flinn, J. Chromatogr., *82*, 117 (1973).
8. J. W. Fairbairn, in *Quantitative Paper and Thin Layer Chro-
    matography*, E. J. Shellard, Ed., Academic Press, London, 1968.

CHAPTER 5

The Mobile Phase

5.1 INTRODUCTION

Success in the technique of thin layer chromatography depends to a great extent upon selecting the mobile phase that will give the desired separation. Much of what is known about selection of a satisfactory mobile phase has been derived from the experiences of many TLC investigators working by trial and error. Work in column chromatography has also been helpful. There has been little work on the theoretical aspects of selecting a sorbent for particular analyses, most of the focus being on mobile phase selection. The nature of the compounds to be separated dictates the sorvent to be used and perhaps even more the selection of the mobile phase.

Few reports describing separations carried out by TLC indicate the rationale behind the choice of mobile phase. It is hoped that the discussions in this chapter will aid the TLC practitioner toward a more organized selection procedure. This should not only save time but also result in better separations. This chapter is divided into sections according to the type of separation, adsorption, ion-exchange, and partition chromatography. In TLC there are other types of sorbents such as layers impregnated with silver nitrate or cellulose layers. Although these are not as widely used, the discussions of the mobile phases herein will give some general guidelines to selecting the mobile phase for use with them also. Selection should not be confused with selectivity. Selectivity is the property of a solvent which causes separation; whereas selection refers to acts of the analyst in use of a particular mobile phase.

5.2   THE SOLVENT PROPERTIES

Before any discussion of choice of a mobile phase can be meaning-
ful, a brief treatment of the problems related to the properties
of the solvents used is appropriate. If the solvents used for
the mobile phase are not pure further complications may result.
A number of general rules follow:

1.   Chloroform or ether contain traces of ethanol as a preservative. It may not be possible to repeat a chromatogram if
this is not removed routinely by distillation or replaced after
distillation to duplicate the original concentration.

2.   Water content of various solvents at different ambient
conditions can vary widely. Herein is one reason that reproducibility of conditions reported in the literature is poor. Many
solvents are hygroscopic and can be affected by humidity. It is
often beneficial to store such solvents over drying agents.

3.   The conditions and duration of storage of the solvent can
affect reproducibility, since solvents can deteriorate. For example, solvents can become contaminated by autooxidation. Storage
in poorly sealed containers can result in contamination by absorption of atmospheric pollutants and water.

4.   Interaction between mobile phase and solute irrespective
of the sorbent can change chromatogram behavior. Advantage can
sometimes be taken of this to give better separation. Impurities
in solvents can also enter such interactions.

5.   Solvents in mixed mobile phases may interact, if solvents
are not pure or vary in composition from batch to batch, the nature of the mixture will vary even if its preparation is similar
from day to day.

6.   Solvent demixing is sometimes seen during development.
This can happen when a small concentration of a strong solvent is
used with a weaker solvent. At the beginning there is a tendency
for preferential adsorption of the stronger solvent, and the front
can consist of 100% weaker solvent. As the sorbent becomes saturated with the stronger solvent, the second front consisting of
the mixture appears. Thus, two solvent fronts are seen on the
layer.

Solvent mixtures should not be used repeatedly. The mobile
phase should be used only once, when it contains a high proportion of a volatile component such as ammonium hydroxide or ether.
Single-solvent mobile phases can be used repeatedly until they
become contaminated. The following give the reasons for this:

1.   Evaporation of the lower boiling component of the solvent
mixture can be caused simply by opening the chamber at the end of
the development. In the case of a benzene-methanol mixture the

methanol concentration gradually lowers on repeated use of the mobile phase.

2. Humidity can change the composition of the solvent mixture when the chamber is opened.

3. Preferential adsorption of one solvent by the sorbent in the layer is often seen. The solvent of highest strength is adsorbed more strongly, and if its proportion is low at the beginning it will eventually affect the composition.

4. The individual components of the mobile phase may react with each other and cause further changes in the real composition.

5. Continued elution of extraneous material from the sorbent can cause changes in the mobile phase composition and produce spurious zones on the layer. This effect is exaggerated the more often a mobile phase is used.

5.3 NATURE OF PHASE INTERACTIONS

There are a number of chemical and physical characteristics of compounds that determine the degree to which a pair can interact with each other. This is true for mobile phase-solute, solute-sorbent and mobile phase-sorbent interactions. The important ones are listed below, they should be kept in mind since the mobile phase-solute interaction controls selectivity and thus separation as discussed later.

1. Intramolecular forces, which hold neutral molecules together in the liquid or solid state are well known. These are physical in character and do not involve formation of chemical bonds. These forces are physical and they are characterized by low equilibrium and result in good chromatographic separation.

2. Inductive forces, which exist when a chemical bond has a permanent electrical field associated with it; for example, C-Cl or $C-NO_2$ groups. Under influence of this field, the electrons of an adjacent atom, group, or molecule are polarized so as to give an induced dipole moment. This is a major contributing factor in the total adsorptive energy on alumina. Thus, the interaction is strong in these cases.

3. Hydrogen bonding makes a strong contribution in adsorptive energies between solutes or solvents having a proton-donor group and a nucleophilic polar surface such as that of alumina or silica gel. These are covered by hydroxyl groups (OH). The hydroxyl groups of silica can also react weakly with electrophilic groups such as ethers, nitriles, or aromatic hydrocarbons to form strong linkages (interactions).

4. Charge transfers between components of the mobile phase and the sorbent can also take place to form a complex of the type S^+A^- (where S = solvent or solute, and A = surface site of sor-

bent). This is prominent in ion-exchange chromatography.

5. Covalent bonds can be formed between solute and/or the mobile phase and the sorbent. These are strong forces and result in poor chromatographic separation.

It should be kept in mind that these factors are present whether mobile phase-solute, mobile phase-sorbent, or sorbent-solute interactions are considered, and furthermore each one of these interactions is independently variable.

5.4 NATURE OF THE SOLUTE

The choice of mobile phase depends on the nature of the compound to be separated. The interaction between the solute-mobile phase or solute-sorbent will *a priori* be determined by the number and nature of the functional groups in the solute. A carboxyl group confers high interactivity to an organic molecule. At the other extreme, the series hydrocarbons with no functional group represent substances of low interactivity. The old rule of thumb applies well here: "like dissolves like." A polar compound will require a mobile phase of high interactivity if it is to migrate on the TLC. Saturated hydrocarbons are the least interactive of the organic compounds, and a mobile phase of lower strength can be used. Unsaturated hydrocarbons are more strongly adsorbed; the more double bonds and conjugation, the greater the adsorption. Introduction of functional groups into the molecule can increase interactivity and thus the adsorptivity or interaction with the sorbent and mobile phase. Functional groups affect the interactivity in increasing degree in the order: RH; $ROCH_3$, $RN-(CH_3)_2$; RCO_2CH_3; RNH_2; ROH: $RCONH_2$; RCO_2H. Thus, knowledge of the nature of the component to be separated provides a starting point for choosing a mobile phase. The above sequence also holds for components of the mobile phase. These properties provided some of the earliest data relating solute parameters to separation efficiency in chromatography. Usually some knowledge of the solubility of the solute is available due to experience with the substance as well as some knowledge of the class into which the compound fits. The degree to which the solute interacts with the mobile phase as opposed to the sorbent will determine its distribution between the two phases. The mobile phase is usually less interactive than the sorbent. If the solute has stronger interaction with the mobile phase, it will prefer the mobile phase and separation will be poor. Conversely, if the solute has a higher interaction with the sorbent, separation will be greater. For some intermediary interactions of the solute with the mobile phase and the sorbent, the distribution will be equal in both phases. These two interplays affect separation efficiency.

Sorbents are usually described by their chemical composition;

that is, they are identified as alumina, ion-exchange, silica gel, or polyamide. There are significant differences in separation selectivity from one sorbent to another. However, there has been no real attempt to deliberately select a particular sorbent for any specific separation. Rather, the composition of the mobile phase is usually altered to produce changes in separation factors. The activity of alumina, for instance, can be changed by varying its water content. This is also true of silica gel. In practice, however, silica gel is usually heated to have high adsorptivity, and separation can be varied by changing the water content.

5.5 THE SOLVENT STRENGTH PARAMETER

The solvent strength parameter $\varepsilon°$ is defined by Snyder (1) as the adsorption energy per unit of standard sorbent. The physical and chemical forces listed in Section 5.3 are involved in determining the level of interaction of the sorbent and mobile phase. The higher the mobile phase strength (greater interaction with the sorbent), the greater will be the R_f of the solute in simple liquid-adsorption TLC. Usually one tries to select a mobile phase such as to obtain R_f values of 0.3-0.7, particularly if well-separated zones are desired. In practice, however, single-solvent mobile phases usually do not give separation although they might give proper mobility. Decreasing solvent strength can increase resolution, but in difficult separations changing mobile phase composition to change interactions may be required. However, at the same time decreasing solvent strength will decrease the R_f and the required separation may not be achieved. Subtle changes in selectivity are the rule. (Selectivity is discussed in later sections.) Mobile phase strength in liquid-solid chromatography is independent of the mobile phase-solute interaction. Strength as defined by Snyder is a function of free energy of adsorption on the solid surface (sorbent). Thus, when a mobile phase has proper strength to give the desired R_f it may not achieve the desired separation.

A starting point for the beginner can be found in some reported mobile phases (based on class of compound) as given in Table 5.1. These mobile phases have been developed for specific separations. A more precise approach to separations based on these mobile phases and how to modify them can be obtained by understanding the theories of mobile phase strength and selectivity. Varying the selectivity (changing interactions) can affect the separation. Solvent strength is dependent on its reactivity with the sorbent. With a large number of sorbents available, one has to resort to some trial and error to find the best mobile phase for separating the components of interest. Activity of the sorbent, particularly in the case of silica gel and alumina, can be

TABLE 5.1   SELECTED SOLVENT-SORBENT SYSTEMS FOR THE TLC
SEPARATION OF VARIOUS COMPOUNDS[1]

---

1. *Chloroplast Pigments*
   a. Isooctane-acetone-ether (3:1:1)--silica gel and polyamide
   b. Petroleum ether-benzene-CHCl$_3$-acetone-isopropanol (50:35:
      10:5:0.17)--cellulose
   c. Petroleum ether-n-propanol (99.2:0.8) followed by 20%
      CHCl$_3$ in petroleum ether (two-dimensional)--sucrose
   d. Petroleum ether-acetone (4:6)--alumina
   e. Petroleum ether-n-propanol (199:1)--starch
2. *2,4-Dinitrophenylhydrazones of Aldehydes and Ketones*
   a. Hexane--ethyl acetate (4:1 or 3:2)--silica gel
   b. Benzene or CHCl$_3$ or ether or benzene-hexane (1:1)--alumin
3. *Alkaloids*
   a. Benzene-ethanol (9:1) or CHCl$_3$-acetone-diethylamine (5:4:
      1)--silical gel
   b. CHCl$_3$ or ethanol or cyclohexane-CHCl$_3$ (3:7) plus 0.05%
      diethylamine--alumina
   c. Benzene-heptane-CHCl$_3$-diethylamine (6:5:1:0.02)--cellulos
      impreg. with formamide
4. *Amines*
   a. Ethanol (95%)-NH$_3$ (25%) (4:1)--silica gel
   b. Acetone-heptane (1:1)--alumina
   c. Acetone-H$_2$O (99:1)--kieselguhr G
5. *Sugars*
   a. Benzene-acetic acid-methanol (1:1:3)--silica gel buffered
      with boric acid
   b. n-Propanol-conc. NH$_3$-H$_2$O (6:2:1)--silica gel G
   c. Butanol-pyridine-H$_2$O (6:4:3) or ethyl acetate-pyridine-
      H$_2$O (2:1:2)--cellulose
   d. Ethyl acetate-isopropanol-H$_2$O (65:24:12 or 5:2:0.5)--
      kieselguhr G buffered with 0.02N sodium acetate
   e. Ethyl acetate-benzene (3:7) (for sugar acetates)--starch-
      bound silicic acid
6. *Carboxylic Acids*
   a. Benzene-methanol-acetic acid (45:8:8)--silica gel
   b. Methanol or ethanol or ether--polyamide
   c. Isopropyl ether-formic acid-H$_2$O (90:7:3)--kieselguhr G-
      polyethylene glycol (M-1000) (2:1)
7. *Sulfonamides*
   a. CHCl$_3$-ethanol-heptane (1:1:1)--silica gel G
8. *Food Dyes*
   a. Methyl ethyl ketone-acetic acid-methanol (40:5:5)--silica
      gel G

TABLE 5.1 (continued)

    b. Butanol-ethanol-H_2O (9:1:1,8:2:1,7:3:3,6:4:4 or 5:5:5)--
        alumina
    c. Aq. sodium citrate (2.5%)-NH_3 (25%) (4:1)--cellulose

. *Essential Oils*
    a. Hexane--starch-bound silicic acid
    b. Benzene-$CHCl_3$ (1:1)--silica gel G

0. *Flavonoids and Coumarins*
    a. Ethyl acetate-Skellysolve B--starch-bound silicic acid
    b. Methanol-H_2O (8:2 or 6:4)--polyamide
    c. Toluene-ethyl formate-formic acid (5:4:1)--silica gel G
        + sodium acetate
    d. Petroleum ether-ethyl acetate (2:1)--silica gel G

1. *Metal Ions*
    a. Dilute HCl--starch-bound alumina--Celite
    b. Acetone-conc. HCl-2,5-hexanedione (100:1:0.5)--silica
        gel G.
    c. 1M aq. $NaNO_3$--Dowex 1 + cellulose
    d. Methanol--alumina

2. *Insecticides*
    a. Cyclohexane-hexane (1:1) or CCl_4-ethyl acetate (8:2)--
        silica gel G
    b. Hexane--alumina
    c. Heptane saturated with acetic acid--starch-bound silicic
        acid
    d. Chloroform--silica gel G + oxalic acid

3. *Lipids*
    a. Petroleum ether-diethyl ether-acetic acid (90:10:1 or
        70:20:4)--silica gel G
    b. Petroleum ether-diethyl ether (95:5)--alumina
    c. $CHCl_3$-methanol-H_2O (80:25:3)--silicic acid

4. *Fatty Acids*
    a. Petroleum ether-diethyl ether-acetic acid (70:30:1 or 2)-
        silica gel G
    b. Acetic acid-CH_3CN (1:1)--kieselguhr impreg. with undecane
    c. Benzene-diethyl ether (75:25 or 1:1)--starch-bound silic-
        ic acid
    d. CH_3CN-acetic acid-H_2O (70:10:25)--silica gel G impreg.
        with silicone oil

5. *Glycerides*
    a. $CHCl_3$-acetic acid (99.5:0.5)--silica gel G impreg. with
        $AgNO_3$
    b. $CHCl_3$-benzene (7:3)--silica gel G
    c. $CHCl_3$-methanol-H_2O (5:15:1)--silica gel G impreg. with
        undecane

TABLE 5.1 (continued)

16. *Glycolipids*
    a. Propanol-12% NH_3 (4:1)--silica gel G
17. *Phospholipids*
    a. $CHCl_3$-methanol-H_2O (60:35:8 or 65:25:4)--silica gel G
18. *Nucleotides*
    a. 0.15M NaCl or 0.01-0.06N CHl--Ecteola cellulose
    b. Sat. aq. $(NH_4)_2SO_4$-1M sodium acetate-2-propanol (80:18:
        2)--cellulose
    c. 0.02-0.04N aq. HCl--DEAE cellulose
    d. Gradient elution:Start with 1N formic acid and add 10N
        formic acid that is 2M in ammonium formate DEAE Sephadex
        A-25
    e. 1.0-1.6M LiCl--cellulose PEI
19. *Phenols*
    a. Xylene, $CHCl_3$ or xylene-$CHCl_3$ (1:1,3:1,1:3)--starch-
        bound silicic acid or silicic acid-kieselguhr (1:1)
    b. Benzene--alumina plus acetic acid
    c. Benzene-1,4-dioxane-acetic acid (90:25:4)--silica gel G
    d. Diethyl ether--alumina
    e. Hexane-ethyl acetate (4:1 or 3:2)--silica gel plus
        oxalic acid
    f. Hexane or cyclohexane or benzene--δ-polycaprolactam
    g. Ethanol-H_2O (8:3) containing 4% boric acid and 2% sodium
        acetate--silica gel G plus boric acid
    h. CCl_4-acetic acid (9:1) or cyclohexane-acetic acid (93:7)--
        polyamide 6
20. *Amino Acids*
    a. Butanol-acetic acid-H_2O (3 or 4:1:1) or phenol-H_2O (75:
        25) or propanol-34% NH_3 (67:33)--silica gel G
    b. Butanol-acetic acid-H_2O (4:1:1)--cellulose
    c. Butanol-acetic acid-H_2O (3:1:1) or pyridine-H_2O (1:1 or
        80:54)--alumina
    d. Ethanol-NH_3 (conc.)-H_2O (7:1:2)--silica gel G buffered
        with equal portions of 0.2M KH_2PO_4 and 0.2M Na_2HPO_4
    e. n-Butanol-acetone-NH_3-H_2O (10:10:5:2) followed by
        isopropanol-formic acid-H_2O (20:1:5) (two-dimensional)--
        cellulose
21. *Polypeptides and Proteins*
    a. $CHCl_3$-methanol or acetone (9:1)--silica gel G
    b. H_2O or 0.05M NH_3--Sephadex G-25
    c. Phosphate buffers--DEAE Sephadex A-25
22. *Steroids and Sterols*
    a. Benzene or benzene-ethyl acetate (9:1 or 2:1)--silica gel
        G
    b. $CHCl_3$-ethanol (96:4)--alumina

TABLE 5.1 (continued)

23. *Terpenoids*
    a. Hexane or hexane-ethyl acetate (85:15)--starch-bound
       silicic acid
    b. Benzene or benzene-petroleum ether or -ethanol mixtures--
       alumina
    c. Isopropyl ether or isopropyl ether-acetone (5:2 or 19:1)-
       silica gel G
24. *Vitamins*
    a. Methanol, CCl_4, xylene, $CHCl_3$ or petroleum ether--alumina
    b. Methanol, propanol, or $CHCl_3$--silica gel G
    c. Acetone-paraffin (H_2O sat.) (9:1)--silica gel G impreg.
       with paraffin
25. *Barbiturates*
    a. $CHCl_3$-n-butanol-25% NH_3 (70:40:5)--silica gel
26. *Digitalis Compounds*
    a. $CHCl_3$-pyridine (6:1)--silica gel
27. *Polycyclic Hydrocarbons*
    a. CCl_4--alumina
28. *Purines*
    a. Acetone-$CHCl_3$-n-butanol-25% NH_3 (3:3:4:1)--silica gel

---

[1] From G. Zweig and S. Sherma, eds., *Handbook of Chromatography*,
C.R.C. Press, Cleveland, 1972, by permission of the Editors.

controlled by controlling its water content. This is probably
the main reason why it has been so difficult to duplicate many
separations reported in the literature. There is at present no
way to state sorbent activity other than to indicate the proportion of water content. The surface activity of the sorbent is
important and can be related to surface area. Modern sorbents for
chromatography are more carefully controlled so separations can
be more reproducible. Stronger mobile phases give greater interaction with the solute and will decrease adsorption, thus causing
faster migrations. Conversely, a weak mobile phase will give a
low R_f.
    An expression of solvent strength can be obtained in the wa-
ter solubility of a variety of solvents. Table 5.2 gives the water solubility of a number of solvents. The squence of solvents
based on increasing or decreasing solubility results in an eluotropic series. Modern eluotropic series are based on solvent
strength parameter ($\varepsilon°$), as will be discussed in following sections.

TABLE 5.2   ELUOTROPIC SERIES BASED ON WATER SOLUBILITY

	Solubility in water (%)
Ethanol	∞
Ethyl acetate	8.60
Butanol	7.90
Ether	7.50
Methyl chloride	2.0
Methyl isobutyl carbinol	1.8
Isopropyl chloride	0.34
Isopropyl ether	0.20
Benzene	0.08
Toluene	0.05
Hexane	0.014

5.6   THE ELUOTROPIC SERIES OF SOLVENTS

In the expression of mobile phase strength the eluotropic series
has been widely used.   As early as 1940, Trappe (2) described a
series of solvents in order of eluting power.   Steiger and
Reichstein used the principle in 1938 in their early work on the
chromatography of adrenal steroids (3).   Jacques and Mathieu (4)
have stated that eluting power of solvents was proportional to
their dielectric constant ε.   They showed that the solvents function by being themselves adsorbed.   In a mobile phase, the solvent
with the greatest strength (highest ε) is the most strongly adsorbed.   These early reports stimulated attempts to systematize
the choosing of mobile phases in chromatography.

A "mixotropic" series of solvents was described by Hecker in
1954 (5).   He noted that the further apart two solvents were from
each other in a series, the less miscible they were with each
other.

It will be noticed that even in the solvent strength or dielectric constant the functional groups of the molecules will follow the sequence given in Section 5.4.

In his work with steroids, Neher (6) applied the principles
of mobile phase selectivity and strength to problems of separation by TLC.   His report may be one of the first real attempts to
use the interplay of mobile phase-sorbent and mobile phase-solute
interactions for the separation of solutes difficult to resolve.

Figure 5.1 shows how a series of mobile phases with different

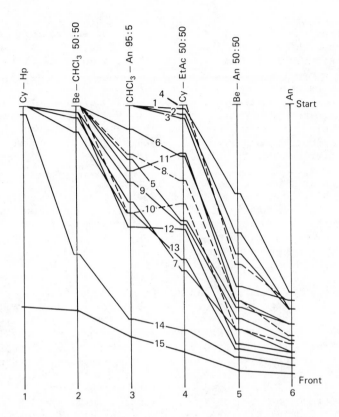

Figure 5.1. Migration distances of sterols and steroids of different polarities in mobile phases with different eluotropic properties. The numbers 1 to 6 (down) correspond to the numbered circles in Figure 5.2. The R_f values in system No. 1 represent the mean values of the very similar values in cyclohexane (Cy) and heptane (Hp); Be = benzene, An = acetone, EtAc = ethyl actate. The numbers 1-15 referring to joint R_f lines represent the steroid as given in the text. This diagram gives a guide to the choice of solvents over a wide range of polarity. Here, the R_f values of a number of sterols and steroids, i.e., from tetrahydrocortisol (No. 1) down to cholestane (No. 15), in six pure and mixed solvent systems of increasing eluotropic power are indicated graphically. The intermediary substances are: (2) cortisol, (3) cortisone, (4) corticosterone, (5) estradiol-17β, (6) 5β-pregnane-3α, 20α-diol, (7) estrone, (8) testosterone, (9) 3β-hydroxy-pregn-5-en-20-one, (10) androst-4-ene-3,17-dione, (11) cortexone, (12) progesterone, (13) cholesterol, and (14) cholesterol acetate.

strengths were used to separate a number of steroids (6). It illustrates the effect of interaction on the separation. Figure
5.2 shows the series of "equieluotropic" solvents that was used
by Neher to select mobile phases of the same strength but of different composition that resulted in changes of interactions and
hence separation.

In the first figure there were only six solvents. However,
there are hundreds of possible and useful mixtures for all types
of steroids or other compounds. This is why the type of "nomo-
gram" illustrated in Figure 5.2 evolved. It was derived from the
mean R_f values for TLC of 20 steroids and a number of dyes. On
the left side of the mixing line is pure solvent of the weaker
strength, and on the right side is the pure solvent of higher
strength. The proportions in between are given on a logarithmic
scale. Pure methanol (high strength) is somewhere out of the
picture to the right. Using this figure, it is possible by trial
and error to formulate a mobile phase that will separate the
steroid or other compound under consideration.

In Figure 5.2, the pure solvents and their binary mixtures,
that is, the mixing lines, are placed empirically under each
other in such a way that the solvents of equal average eluting
power for the tested steroids are to be found in the vertical dimension. Thus, the ability to move solutes increases from left
to right, whereas it remains practically constant along the vertical lines. The numbered circles (in Figure 5.2) 1-6 (from left
to right) refer to the composition of the mobile phases of increasing strength shown in Figure 5.1. On the other hand, the
vertical line X has been drawn through the binary mixing lines in
an arbitrary position. If, for example, estradiol-17β and cholesterol were developed together in the mobile phase whose composition is given by this line X, the R_f values of each steroid would
remain more or less constant in the various mobile phases, but
because of mobile phase-solute interaction different mobilities
can result.

The results in practice are shown in Figure 5.3. At the top
are listed the mobile phases composed according to line X of Figure 5.2. On the lines between start and front, the experimental
R_f values of (1) testosterone, (2) estradiol-17β, and (3) cholesterol have been marked. The R_f values fluctuate about a certain
mean value.

As far as "strength" is concerned, not many mobile phases
are needed to cover the whole range of steroid chromatography; but
to separate all the mixtures that are encountered, it is advisable
to use several mobile phases of approximately equivalent strength,
as may be seen from the intersecting zig-zag lines in Figure 5.3.
This means that if a steroid mixture cannot be separated in one
mobile phase of suitable strength, a mobile phase of the same

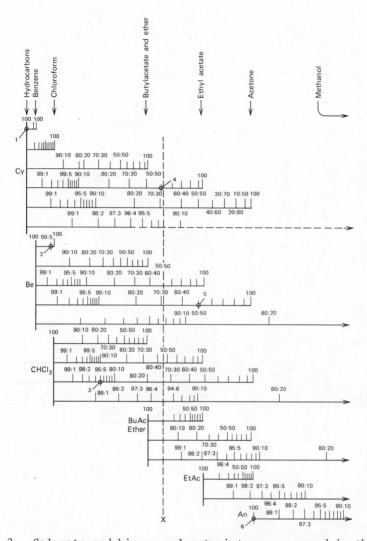

Figure 5.2.   Solvents and binary solvent mixtures arranged in the "equieluotropic series" on the basis of average R_f values for 20 steroids.   Each vertical line connects solvent mixtures of the same average eluotropic properties; pure methanol is somewhere out of the figure to the right.   The first line section from the left or right in the "mixing line" joining two solvents always represents a mixture of 90% and 99%, respectively, of the nearest pure solvent, and 10% or 1% of the other, etc.

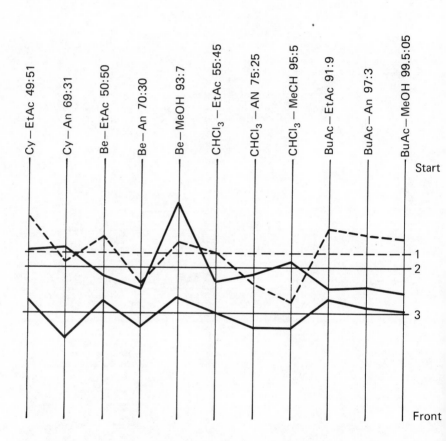

Figure 5.3. R_f values of (1) testosterone, (2) estradiol-17β, and (3) cholesterol in the binary solvent mixtures given by the vertical line X in Figure 5.2. The R_f values of each steroid vary within a certain "equieluotropic" range about the mean value (straight lines). An = acetone; Be = benzene; BuAc = n-butyl acetate; Cy = cyclohexane; EtAc = ethyl acetate; MeOH = methanol.

strength but of completely different composition may be required.
This system of equieluotropic solvent mixtures is, of course, also
applicable to column chromatography. The principle is based on
the fact that different solutes interact differently with the different solvent components in the mobile phase. If two solutes do
not separate in a given system, changing to a mobile phase of the
same strength that interacts differently with the solutes will
often increase the resolution.

Mobile phase strength affects mobility. As a general rule
decreasing strength increases resolution. With a large number of
solvents available there is an infinite number of mobile phase combinations possible. Therefore, it is appropriate to consider selectivity or interactions between the solute and mobile phase.
This is discussed in detail in the following sections.

5.7 SOLVENT SELECTIVITY

The organization of mobile phase selection parameters is based on
the work of Snyder, which is summarized in his widely used monograph (1). The expression of resolution resulting from this and
other work can prove very useful in choosing an appropriate
mobile phase in both column and thin layer liquid chromatography.
Snyder's mathematical expressions of adsorption chromatography
have made it possible to systematize choice of mobile phase. Further advances in the theory were described in his report published
in 1971 (7). The basic books on liquid chromatography (8-10) contain complete discussions of the theories of adsorption and interactions between the mobile and stationary phases and the solute.

Solvent selectivity refers to the ability of a solvent to
create differences in the relative migration of two solutes. The
general formula for separation power or resolution (Rs) is expressed in the equation:

$$Rs = \frac{1}{4} (\alpha - 1) \sqrt{N} \left[\frac{k'}{k' + 1} \right]$$

The separation factor α is the ratio of capacity factors k_1/k_2
for two solutes. N is the number of theoretical plates in the
adsorbent bed through which the mobile phase flows and through
which the solute passes. k' is the average of k_1 and k_2 (capacity
factor k is the equilibrium ratio of total solute in the stationary phase to total solute in the mobile phase). Thus, resolution, a separation power, is dependent on three factors: N is
largely a function of the nature of the sorbent whether in a column or in a thin layer. k' reflects the average migration speed
of the two solutes and is determined by mobile phase strength ε°.

k' and $\varepsilon°$ must be held between narrow limits for optimum separation to be possible; α is determined by the sorbent and the composition of the mobile phase. In thin layer chromatography little attempt has been made to evaluate N; consequently, emphasis has been placed on choice of mobile phase. Selectivity is largely based on differences in interaction between the mobile phase and solute. Different solutes will interact differently with the mobile phase. Therefore, if two solutes cannot be separated with a mobile phase of predetermined strength, they can sometimes be separated by using a different mobile phase of the same strength. Although mobility is not greatly changed the resolution will be improved.

In practice the application of Snyder's theories require the assembly of a number of sample, sorbent, and mobile phase parameters. Saunders (11) has condensed some of the basic concepts of adsorption activity, group adsorption strength, and mobile phase strength into two graphic forms. The use of the two charts presented here as Figures 5.1 and 5.2 is very simple and rapid and gives a first approximation to a mobile phase appropriate for a given sample. The results are approximate and in a few cases may be in error. However, the technique has been found to be a very favorable alternative to tedious calculations and the trial-and-error approach. Since the purpose of this text is to emphasize practice rather than theory, and Saunders' method is applicable to TLC, we shall describe that method.

The sample partition ratio k' is defined as the ratio of the amount of solute in the stationary phase to the amount in the mobile phase (see previous formula from Snyder). This value can also be calculated from the equation:

$$k' \equiv \frac{t_r - t_0}{t_0}$$

where t_r is the sample retention time and t_0 is the retention time of an unadsorbed component.

Since R_f or mobility factors in TLC also depend on the ratio of solute in the mobile phase to that in the stationary phase, a valid practical use can be made of the expressions that follow: For a given sample and sorbent, log k' varies linearly with $\varepsilon°$. Saunders in his example defines the E_3 value of a sample as the solvent strength required to give k' = 3, as shown in Figure 5.4.

For good results the k' value should be less than 10. With k' of 10 the solute becomes diffused in the sorbent; the separation is poor and because of a long retention time the R_f is low. A careful evaluation of the equation as related to retention times and the limitation of k' to between 1 and 10 indicates that these limits correspond relatively close to the limits of R_f 0.3-0.7 proposed in this book and by other authors as the region of useful separation in TLC. A value for k' of 1.0 in columns corresponds roughly to an R_f of 0.1 in TLC. Thus, a k' value of 3 is a good working point corresponding to an R_f between 0.3 and 0.5. The essential point is to keep the R_f above 0.1 and less than 0.7. Thus, the following is applicable to TLC as well as columns.

Figure 5.4 shows the E_3 values for a variety of compounds on a typical silica. The following describes how to approximately determine the solvent strength for a very broad variety of mono- and polyfunctional compounds and how to obtain several alternative mobile phases for any solvent strength using only the two simple graphs. Consider naphthalene as a hydrocarbon (second bar from top) in Figure 5.4. The range of E_3 values extends from 0.09 for fully activated adsorbent (equilibrated with dry solvent) down to -0.13 for highly deactivated adsorbent. This means that with an active silica and a solvent strength of 0.09, naphthalene will separate with k' = 3. With deactivated silica, weaker solvents will be needed to give k'=3 down to a lower limit of about -0.13 for highly deactivated silica.

By grouping various parameters and comparing their relative effects on E_3, it is possible to formulate the following approximate rules for estimating E_3 in a broad variety of polyfunctional compounds. In the examples accompanying these rules, E_3 values cited are for fully activated silica.

1. Saturated hydrocarbons and simple olefins are eluted with k'=3 with all solvents. Fluorinated solvents are an exception because of lower solvent strength.

2. Alkyl- and halogen-substituted aromatics have E_3 values within ±0.05 unit of the parent compound.

3. Aliphatic compounds with 2-20 carbons will have E_3 within ±0.05 unit of the model compound (R = C_6H_{13}) used in Figure 5.4.

4. Difunctional compounds will have E_3 values larger than either of the monosubstituted compounds. For example, for benzaldehyde E_3 = 0.23 and for methoxybenzene E_3 = 0.15, but the E_3 value for methoxybenzaldehyde will be larger than 0.23 by an increment that depends upon the difference between E_3 values (ΔE_3) of the monosubstituted compounds as follows:

ΔE_3	Increment
0-0.1	0.15
0.1-0.2	0.07
0.2	0.00

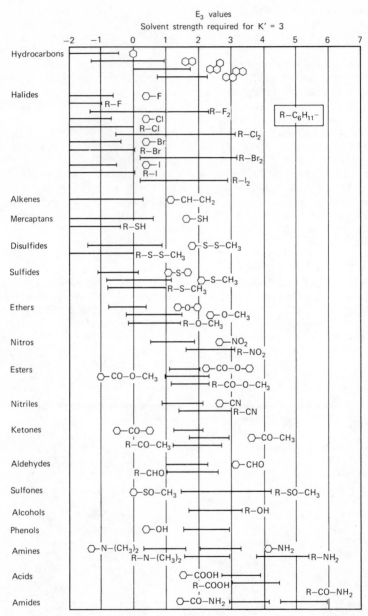

Figure 5.4. E_3 values of various compounds on silica gel. See text for nature of E_3.

In our example E_3 = 0.23 - 0.15 = 0.08; thus for methoxybenzaldehyde, the increment is 0.15 and $E_3 \tilde{} $ 0.23 + 0.15 = 0.38. Similar-
ly, for diaminohexane, ΔE_3 = 0.0, $E_3 \tilde{} $ 0.54 + 0.15 = 0.69.
   5. Additional functional groups will add successively
smaller increments to E_3, but the size of these increments cannot
be easily predicted.
   6. Addition of an aliphatic halogen or mercaptan moiety will
increase E_3 by an increment 50-100% larger than the value indicated in 4 and 5 above. For example, for chlorohexyl ethyl ether
$E_3 \tilde{} $ 0.15 + 2(0.07) = 0.29.
   7. Compounds containing polynuclear aromatic moieties
should be considered benzene compounds substituted with additional
aromatic rings. For example, nitrophenanthrene should be considered a nitrobenzene (E_3 = 0.19) substituted with two additional
rings, that is, naphthalene (E_3 = 0.09). Therefore, $\Delta E_3 \tilde{} $ 0.1
and the corresponding increment is 0.15; thus, E_3 = 0.34. Once
the E_3 values have been estimated from Figure 5.4 and the preceding rules, we can choose an appropriate solvent mixture from Figure 5.5 The six solvents used in Figure 5.5 are among the most
useful for adsorption chromatography. Viscosities are low, allowing maximum resolution per unit time, and the whole range of solvent strengths is covered.

   In Figure 5.5, solvent strength ($\epsilon°$) is plotted across the
top versus various binary solvent compositions, in a manner similar to that of Neher in Figure 5.2. Each horizontal line corresponds to a range (0-100% volume) of binary solvent mixtures.
The first five lines represent mixtures of pentane with other solvents. The top line of this series of five corresponds to mixtures of pentane and isopropyl chloride. For any solvent strength
intermediate between pentane and isopropyl chloride ($\epsilon°$ = 0.22),
the required solvent composition is found by dropping a vertical
line from the $\epsilon°$ scale. Thus, for $\epsilon°$ = 0.10 the mixture is 26
vol-% isopropyl chloride in pentane, and for $\epsilon°$ = 0.20 it is 80
vol-% isopropyl chloride in pentane. The next four lines in this
series correspond to mixtures of pentane with methylene chloride
($\epsilon°$ = 0.32), ethyl ether ($\epsilon°$ = 0.38), acetonitrile ($\epsilon°$ = 0.50),
and methanol ($\epsilon°$ = 0.73). The second series of lines correspond
to binary mixtures of isopropyl chloride and a solvent of higher
$\epsilon°$. For any given solvent strength, Figure 5.5 indicates several
binary mixtures. Consider solvent mixtures with $\epsilon°$ = 0.30. One
of these mixtures is 76 vol-% methylene chloride in pentane. Sim-
ilarly, 49 vol-% ethyl ether in pentane or 37 vol-% ethyl ether
in isopropyl chloride, etc., would also give $\epsilon°$ = 0.30. Any of
the indicated solvent mixtures would be an appropriate first
choice for a sample with E_3 = 0.30.

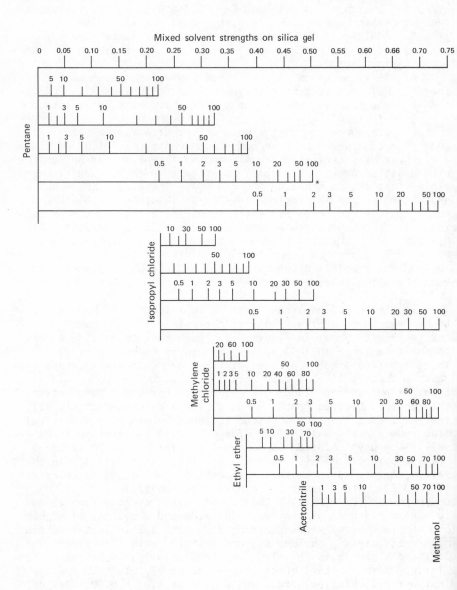

Figure 5.5.   Mixed solvent strengths on silica gel.   See text for
details.

Because of the assumptions and approximations used in the derivation of Figures 5.4 and 5.5 and Rules 1-7, and the limitations of the basic theory, the initial choice of solvent is seldom exactly correct. However, it is usually close enough to allow a rapid adjustment to the optimum. It has been found that the following rules are helpful.

1.   An increase in solvent strength (ϵ°) will decrease separation and increase the R_f.

2.   A decrease in ϵ° will increase separation and decrease R_f.

3.   Selectivity will be greatest if the concentration of the stronger component of the mobile phase is < 5 vol-% or > 50 vol-%. For example, at ϵ° = 0.30, either 76 vol-% methylene chloride in pentane or 1.7 vol-% acetonitrile in isopropyl chloride would be expected to give maximum selectivity due to more favorable solute-mobile phase interactions.

4.   Substitution of ethyl ether or methanol for one of the other strong components of the mobile phase can often improve selectivity due to the formation of hydrogen bonds (9), a potential of these solvents.

5.   If a dry solvent results in tailed spots, a more highly deactivated sorbent should be used to decrease the solute-sorbent interactions. This is usually done by adding water to the sorbent.

For the extension of Figure 5.4 to more complex molecules, Rule 1 is simply a statement of the fact that nonaromatic hydrocarbons are not adsorbed with any of the solvents listed in Figure 5.5. Fluoroalkanes and fluoroethers have solvent strengths less than 0, resulting in some adsorption of olefins and saturated hydrocarbons (6). However, these solvents have found little use in adsorption chromatography, presumably because of their high cost.

Rule 2 and 3 arise from the very small group adsorption energies of alkyl groups generally and of halogens substituted on aromatics.

The graphical and rule-of-thumb approach presented here is an approximate but convenient method for quickly finding the mobile phase appropriate for the separation of a given mixture. Those who wish more accurate prediction should refer to the detailed calculations presented in Snyder's monograph (1).

The mobile phase strength data given in Tables 5.3 and 5.4 are based on results obtained with alumina. These data will be helpful in selection of a mobile phase. Table 5.3 gives solvent strength of binary mixtures of pentane with any of eight different solvents. Table 5.4 gives the data for benzene mixed with other

TABLE 5.3. SOLVENT STRENGTHS FOR MOBILE PHASES OF POLAR SOLVENTS IN PENTANE*

Solvent A: Pentane

$\varepsilon°$	CS_2	i-PrCl	Benzene	Ethyl ether	$CHCl_3$	CH_3Cl_2	Acetone	Methyl acetate
Solvent B:								
0.00	0	0	0	0	0	0	0	0
0.05	18	8	3.5	4	2	1.5	1.5	
0.10	48	19	8	9	5	4	3.5	2
0.15	100	34	16	15	9	8	6	3.5
0.20		52	28	25	15	13	9	5
0.25		77	49	38	25	22	13	8
0.30			83	55	40	34	19	13
0.35				81	65	54	28	19
0.40					100	84	42	29
0.45							61	44
0.50							92	65
0.55								
0.80								100

* Values refer to volume percent B in solvent A (pentane) for the given $\varepsilon°$ value. These data are for alumina as sorbent. For silica gel the order will be the same, but the actual $\varepsilon°$ value will be slightly different.

122

TABLE 5.4. SOLVENT STRENGTHS FOR MOBILE PHASES OF POLAR SOLVENTS IN BENZENE*

	Solvent A.	Pentane	Benzene				
ε°	Solvent B:	Diethyl-amine	Acetone	Methyl acetate	Diethyl-amine	Aceto-nitrile	i-PrOH†	MeOH§
0.30		2.5						
0.35		5	6		2			
0.40		8	18	4	7			
0.45		13	36	12	14			
0.50		22	60	24	26	1.5		
0.55		38	93	42	45	6		
0.60		73		66	77	14	4	
0.65				100		36	7	
0.70					100	100	12	
0.75							21	4
0.80							40	8
0.85							75	18
0.90								44
0.95								100

* Values refer to volume percent of solvent B in solvent A (benzene) for the given ε° value. These data are for alumina as sorbent. For silica gel the order will be similar but ε° will be different.
† Isopropanol
§ Methanol

solvents. Using these tables one can find different mobile phases
with the same strength but different selectivities by simply following a line through the point of required mobile phase strength
($\varepsilon°$).

To summarize solvent selection: If the R_f is too high,
choose a mobile phase of lower strength ($\varepsilon°$). If it is too low,
go to a higher $\varepsilon°$. Once the desired mobility is attained, work
on selectivity. If a pair does not separate with the mobile phase
(mixture of solvents) being used, select a mobile phase with the
same strength but different components and thus different interactions. The tables included in this chapter provide a number of
mobile phases of the same strength but of widely varying compositions, which will provide different solute-mobile phase interactions.

5.8   ION EXCHANGE

Ion exchange chromatography is widely used in the separation of
amino acids, proteins, protein hydrolysates, and other ionized
compounds. The theory for selection of solvents is not as well
advanced as in adsorption chromatography. However, advances in
resin technology in the past few years have resulted in the introduction of some very useful ion-exchange sorbents. This technique is more often used in column chromatography. A number of
ion exchangers are available for thin layer separations (see Chapter 2 on sorbents).

Aqueous mobile phases containing various ionic components are
used for ion-exchange chromatography, with resolution and solute
mobility varied by changes in pH and ionic strength. Small
amounts of organic solvent are added to increase mobility of
strongly retained compounds as well as to increase solubility.
In ion-exchange TLC, the regeneration or activation of the layers
and development of the layer must be distinguished. Conversion to
the necessary ion form is usually carried out before the sample is
applied to the plates. It can also be performed on the plate with
the solvent or buffer used to develop the chromatograms. In many
cases it is sufficient to develop the layer with water or 0.1N
sodium chloride solution. This type of pretreatment of the plate
often leads to better separation.

There are a few rather basic precepts learned from column
chromatography in terms of the mobile phase. The mobile phases
used can be divided into four classes:

1. Buffer solutions of various ionic strengths within the
pH range 3.0-8.0; for example, 0.005M phosphate buffer of pH 8.0
or 2M sodium formate buffer of pH 3.4.

2. Salt solutions, for example, 0.1-2.0M LiCl or NaCl.
3. Distilled water.
4. Any one of the above with organic solvent added.

Table 5.5 shows some buffers and pH ranges typically used to separate mixtures of amino acids on cation exchangers. Several different buffers are usually used for each analysis.

TABLE 5.5. BUFFERS FOR USE WITH CATION EXCHANGE

	Protein hydrolysate	Physiologic fluid
Composition	Sodium citrate	Sodium or lithium citrate
Concentration	0.2-1.5M sodium	0.2-1.5M sodium or 0.3-0.8M lithium
	0.03-0.1M citrate	0.05-0.20M citrate
pH	2.8-6.0	2.5-6.0

A large number of other buffers or salt solutions can be used, depending on the nature of the compound to be separated--Is it acidic or basic? Is it a cation or an anion? In a general sense, pH will be controlled by the use of buffer, and the ionic strength can be adjusted by the addition of some other salt. An example of this is the borate buffer of pH 9.2 with ionic strength adjusted by adding 0.002N sodium nitrate. The selectivity is kept in narrow limits by controlling the ionic strength.

A basic understanding of the process taking place in the ion-exchange phenomenon will facilitate selection of the mobile phase for use with sorbents for ion exchange that have charge-bearing functional groups. The most common mechanism is simple ion exchange as expressed in the following formulas:

$$S^- + M^+Y \longleftrightarrow Y^- + R^+X^- \quad \text{(anion exchange)}$$

$$S^+ + M^-Y^+ \longleftrightarrow Y^+ + R^-X^+ \quad \text{(cation exchange)}$$

Where S = sample ion
 M = mobile phase ion
 R = ionic site on exchanger
 X = sample anion
 Y = sample cation
   In anion exchange, the sample anion S^- is in competition with
the mobile phase ion M^- for ionic sites on the exchanger, R^+. Conversely, in cation exchange, sample cation S^+ is in competition
with the mobile phase cation M^+ for sites on the exchanger, R^-.
Simple ion-exchange separations are based on the different
strengths of the solute ion/resin interactions. Solutes that
interact weakly with the exchanger in the presence of the mobile
phase are poorly retained on the sorbent and show higher mobility.
Conversely, solutes that interact strongly are strongly retained
and move more slowly in a layer or column.
   Both acids and bases can be separated by ion exchange. A
weak acid, HA, dissociates into its ionic form, H^+ and A^-:

$$HA \longleftrightarrow H^+ + A^-$$

The degree of this dissociation can be controlled by the pH of
the system. By increasing the concentration of hydrogen ions (decreasing the pH), the dissociation equilibrium can be forced to
the left, thus decreasing the concentration of A^- ions available
for interaction with anion-exchange resin and decreasing the k'
value for solute HA. Change in pH can also control the availability of ionic forms of bases for ion-exchange separations:

$$B + H^+ \longleftrightarrow BH^+$$

In this case, increasing the concentration of hydrogen ions (decreasing the pH) shifts equilibrium to the right, thereby increasing the concentration of the charged (BH^+) ions available for ion-
exchange interaction and increasing the separation of basic solute B.
   Ion exchangers in the acid or base form can also be used to
carry out separations via *acid-base reactions*:
   1. $HA + R^+OH^- \longleftrightarrow R^+A^- + H_2O$
   2. $B + R^-H^+ \longleftrightarrow R^- + BH^+$
where R = ion-exchange resin.
   This technique is useful for separating weak acids and bases
in nonaqueous systems. A weak acid, HA, undergoes interaction
with the anion exchanger, R^+OH^-, to form the resin anion salt,
R^+A^-, plus water. This is a reversible acid-base reaction. The
similar reaction of a weak base, B, reacting with the cationic
resin, R^-H^+, to form a salt, R^-BH^+ is also seen.

5.9 EFFECT OF IONS

Most ion-exchange chromatographic separations are carried out in
aqueous media because of the desirable solvent and ionizing properties of water. Mixed solvents, such as water-alcohol, and even
totally organic solvents have also been used. In ion-exchange
chromatography with aqueous mobile phases, mobility is controlled
by the total salt concentration (or ionic strength) and by the pH
of the mobile phase. An increase in the concentration of the salt
in the mobile phase increases the mobility of solutes. This is
due to the decreased ability of the solute ions to compete with
the mobile phase counterions for the ion-exchange groups in the
resin.

The type of ions in the mobile phase can significantly affect
the mobility of solute molecules, due to the varying interaction
of these different mobile phase ions with the ion-exchange resin.
The retention sequence of various anions for conventional cross-
linked polystyrene anion-exchange resin is as follows:

$$\text{citrate} > SO_4^{2-} > \text{oxalate} > I^- > NO_3^- > CrO_4^{2-}$$
$$Br^- > SCN > Cl^- > \text{formate} > \text{acetate} > OH^- > F^-$$

This retention sequence varies somewhat for different resins, but
it is a qualitative indication of the ability of various anions
to interact with strong anion exchangers. In this sequence, citrate is very strongly bonded to the resin, while fluoride ions
are very weakly bound. Thus, solute molecules will have higher
R_f rates with citrate ions than with fluoride ions.

For many years, when sodium citrate buffers were used almost
exclusively as the mobile phase, a complete analysis of physiolo-
gic fluids by ion-exchange chromatography was hampered by an in-
ability to separate asparagine and glutamine from the other amino
acids (12,13). Either two separate analyses under different conditions were required to separate all of the acids, or two analyses were carried out under the same conditions, with one or two
of the amino acids selectively destroyed between runs (14) and a
difference of peak areas used as a means of quantitation. The
use of lithium citrate buffers instead of sodium citrate has now
made the complete separation of these two important amino acids
possible (15). In this case, the lithium ion is not bound as
strongly to the resin, yielding generally larger k' values. This,
in addition to specific characteristics of the amino acids, gives
some separation advantages with asparagine, glutamine, and the
other amino acids in that region of the chromatogram.

There are some general guidelines for predicting the ability
of an exchanger to complex with solute or mobile phase ions. The
ion exchanger tends to prefer:

1. The ion of higher valence (charge).

2. The ion with a smaller (solvated) equivalent volume.
3. The ion with the greater polarizability.
4. The ion that interacts more strongly with the fixed ionic groups or with the matrix.
5. The ion that interacts weakest with other materials in the mobile phase.

Effect of pH separation in ion-exchange chromatography can also be controlled by changes in mobile phase pH. Thus, an increase in pH decreases mobility in cation-exchange chromatography and increases it in anion exchange. The influence of pH on retention can be explained by the effect of hydrogen ion concentration on available sample ions.

1. $RCOOH \longleftrightarrow COO^- + H^+$ (anion exchange)
2. $RNH + H^+ \longleftrightarrow NH_2^+$ (cation exchange)

Here it is seen that a free carboxylic acid is in equilibrium with the carboxylic ion plus a proton. At highest pH (lower concentration of H^+) the concentration of carboxylic ions is higher; thus, the retention of this sample ion is greater because of its increased ability to compete for the anionic sites of the resin. The formation of the charged cationic form of a weak base is enhanced at higher hydrogen ion concentrations. Thus, a decrease in pH results in lower R_f due to the increased competition of these charged sample ions for the cation-exchange sites. Table 5.6 is from the classical study by Hamilton (12) and shows the effect of changes in pH on the elution order of amino acids. These data illustrate why changes in buffer pH are often used to control selectivity. However, changes in pH can also cause changes in the order of separation of some amino acids.

Variation in pH can also cause changes in the *relative* retention of sample components. Therefore, changes in pH can not only be used to control mobility but can also be useful in changing selectivity. Changes in selectivity as a function of pH are difficult to predict and usually have to be found by a trial-and-error process. When using anion-exchange resins, pH is normally controlled by using cationic buffers (e.g., buffers containing ammonia, pyridine, etc.); with cation-exchange resins, buffers containing ions such as acetate, formate, and citrate often are used to fix the pH in the system. However, other factors can cause variation in pH. It is well known that temperature changes can cause changes in the pH of Tris buffer systems.

With the basic amino acids, an increase in cation concentration improves resolution (band width) and decreases k' value. However, changes in k' are nearly parallel, so these variables are not of much help in selectively moving one acid with respect to the others.

TABLE 5.6   SEPARATION OF ACIDIC AMINO ACIDS (12)

Order	At pH 4.15	At pH 4.80	At pH 5.25
1	Tyrosine	Tyrosine	β-Alanine
2	Phenylalanine	β-Alanine	β-Aminoisobutyric acid
3	β-Alanine	β-Aminoisobutyric acid	Tyrosine
4	β-Aminoisobutyric acid	Phenylalanine	Phenylalanine

5.10   EFFECT OF ADDITION OF ORGANIC SOLVENT

The differential solubility of threonine and serine in various organic solvents (methanol, thiodiglycol, methyl Cellosolve) is an aid in separating them. The addition of up to 10% of one of these organics in the buffer makes a dramatic improvement in the separation of these amino acids. It has been observed (15,16), however, that the addition of these solvents has some effect on the elution pattern of other sample components. All other conditions remaining the same, several amino acids (methionine sulfone, glutamic acid, proline, and citrulline) are eluted sooner in the presence of an organic, and others (glycine, alanine, and α-amino-n-butyric acid) are delayed.

The solvents used in the mobile phase can also provide changes in selectivity. As previously indicated, an aqueous carrier is normally used in ion-exchange chromatography. However, small amounts of organic solvents, such as ethanol, tetrahydrofuran, and acetonitrile, are sometimes added to these aqueous systems to increase the solubility of sample components, and these solvents also can provide changes in selectivity. The addition of organic solvents to aqueous buffers sometimes decreases the tendency of certain sample components to form tailing peaks. Weak acids and bases can sometimes be separated with conventional ion-exchange resins, using nonaqueous carrier systems. In this form of chromatography, decreasing the polarity of the organic decreases the mobility of sample components so that resolution is affected. Changing from methanol to benzene, then to pentane, greatly decreases the mobility of sample components with conventional macroreticular ion-exchange resins (17).

The foregoing points out the problems and some solutions to ion-exchange chromatography. However, it should be remembered that numerous ionizable compounds have also been separated by adsorption chromatography as previously discussed. The solvent systems used must include water, and in many cases acid or basic solvents are also included. Most of the theories of solvent selection in chromatography have been formulated using column chromatography. Therefore, the rules that have been presented may vary slightly when applied to TLC. Many times the investigator has done preliminary evaluations on TLC before going to high-pressure liquid chromatography and can benefit from this knowledge.

The separation scientist is often confronted with a large number of possible systems from which to make a choice for a particular separation. In TLC, qualitative analysis is often performed by using several different solvents in parallel chromatograms. The question arises as to which combination will provide the most information.

The problem in TLC is the choice of a number of mobile phases or sorbents or combination of them so that the separation factors will be as different as possible. These problems are common, but rarely are they stated explicitly, and there are no methods for mathematical classification of chromatographic systems. Often the mobile phases referred to in the literature must be improved to fit the situation at hand.

More recently other attempts to place solvent selection on a firmer basis have been reported. Numerical analysis was used by Turina and Trbojevic (18). Using equations based on R_S values, where R_S is the ratio between the distance of two spot centers according to Giddings (19), they were able to calculate the optimal ratio of components in a TLC system. Numerical analysis was used, and three experiments were required for a two-component system. A three-component system required five to nine experiments. Turina and Trbojevic concluded that the method may be more useful for correcting commonly applied solvent systems than in developing new systems.

Massart and DeClerq (20) classified systems according to mutual resemblances by numerical taxonomy techniques. They illustrated the use of the technique by selecting a combination of three solvent/stationary phases from a set of ten for the separation of 26 different dyes used in the food industry. Those who are interested in improving the selectivity of solvents as well as in more scientific solvent selection will find these two articles helpful.

5.11 LIQUID-LIQUID (PARTITION) TLC

The selection of solvents for liquid-liquid chromatography or partition chromatography is still largely empirical (21). There are no universal partitioning systems for classes of solutes, but

there are infinite capabilities for separation simply by selecting
the appropriate pair of partitionary liquids. Since this type of
thin layer chromatography has seen relatively little use it will
not be discussed at length here, but a few basic generalizations
will be presented. The interested reader should refer to the discussions in the section on liquid-liquid chromatography in the
book *Modern Liquid Chromatography* (21). Some of the generalizations in paper chromatography are applicable. Partition chromatography is often used in the paper chromatography of steroids
using the system as described by Zaffaroni, et al. (22), with propylene glycol as the stationary phase and toluene or benzene as
the mobile phase. The stationary phase can be sprayed on the
plate before using or mixed with the slurry before preparing the
layer. Many partition systems are described for a wide variety
of compounds in the book *Paper Chromatography* (23). Some of these
systems have been adapted to thin layer as well as column chromatography.

In liquid-liquid partition the mobile phase must be immisci-
ble with the stationary phase. Both partitioning phases, but especially the mobile phase, should have as low a viscosity as possible for higher sorbent permeability. Most important is the selectivity of a liquid-liquid system for a given solute.

In partition TLC the resolution of solutes is generally controlled by changing the mobile phase. While it is possible to
vary separation somewhat by selection of the stationary phase,
this is often inconvenient in actual operation. The interaction
of the two liquid phases (mobile and stationary phase) controls
the overall separation for solutes that can be separated in the
particular system. Thus, polar compounds may be separated using
polar stationary phases and nonpolar mobile phases. Conversely,
nonpolar solutes are best separated with a system consisting of a
nonpolar stationary phase and a polar mobile phase. However, once
the selection of the basic system is made, adjustment separation
is normally carried out by varying the mobile phase composition.

In partition TLC, the sorbent acts merely as a support for
the stationary phase. Kieselguhr is often used because of its
low activity. Cellulose is often used. The sorbent can be impregnated in a number of ways depending on the type of partition
system to be used. The plate can be equilibrated with the vapor
of a solvent mixture consisting of the stationary phase saturated
with the mobile phase, much as in paper chromatography. Or the
plate can be sprayed with the stationary phase alone or in a volatile solvent. The plate can also be dipped in the solution, or
the solution can be allowed to migrate up the plate. In each
method the plate must be dried of solvent before application of
the sample. The mobile phase must be saturated with the stationary phase before use. In most cases the stationary phase is high-

ly polar (water, formamide, or propylene glycol) and the mobile
phase is relatively nonpolar.

In reversed-phase partition the stationary phase is nonpolar
(paraffins) and the mobile phase is polar, i.e., water or alcohol
or mixtures of these. In this case it is the nonpolar compounds
that will remain near the origin.

5.12 THE MOBILE PHASE IN GEL THIN LAYER CHROMATOGRAPHY

Gel chromatography can be used for separating higher molecular
species, particularly nonionic molecules such as proteins and nucleic acids. Mixtures of components of different molecular
weights are routinely separated with this method. Hruska and
Franek (24) used Tris buffer (0.2M, pH 8.0) to separate individual albumins on layers of Sephadex G-200 or Sephadex G-75. The
separations were carried out in a moist chamber. Tween 80 (60 ml
per 100 ml buffer) was added to reduce the adsorption of the protein to the gel. The method gave chromatograms that could be
scanned densitometrically as well as by autoradiography.

Blessing and Gebele (25) used Sephadex-100, 150, and 200 in
thin layers for the separation of myoglobin and hemoglobin using
phosphate buffers as the mobile phase.

In gel chromatography, the mobile phase is *not* varied to control resolution. Rather, the mobile phase is chosen for low
viscosity at the temperature of separation and for its ability to
dissolve the sample. The effect of the mobile phase on various
nonrigid sorbents for gel chromatography must be determined. Thus,
in gel filtration, a salt should be added to the solvent to maintain constant ionic strength. Otherwise the passage of ionic samples through the bed can lead to shrinkage of the gel and a decrease in permeability.

5.13 SUMMARY

This chapter shows that TLC can utilize a large number of separation parameters depending on the sorbent of choice. As further
knowledge is gained specific to sorbent-mobile phase, sorbent-
solute, and mobile phase-solute interactions, it may soon be possible to finally systematize mobile phase selection and eventually
to remove the trial-and-error methodology that has plagued the
separation scientist using chromatography as a tool. In TLC, sorbent characteristics are rather constant as opposed to the mobile
phase. This is the reason so much emphasis is placed on these
characteristics. Perhaps the use of liquid-liquid chromatography
or partition in TLC will increase and further the utility of this

technique in separations. However, as pointed out earlier, many investigators first assess the applicability of a system by means of TLC.

If the sorbent is to be deactivated with water, dry solvents cannot be used. However, most TLC is done with plates that are activated by heating just before use, and it is necessary to use dry solvents to prevent deactivating the sorbent. Thus, the condition of the mobile phase itself can determine the effectiveness of the separation. If activated plates are used, attention must be given to the amount of water in the mobile phase. Different solvents will have different water content. Therefore, it is necessary to bear these factors in mind when reporting the mobile phase and the activity of the sorbent used in a specific separation. Some chromatographers use solvents containing a known amount of water. The sorbent will dry the solvent as it moves up the plate. The sorbent then is somewhat deactivated, and an equilibrium condition is set up that sometimes improves the separations.

REFERENCES

1. L. R. Snyder, *Principles of Adsorption Chromatography*, Marcel Dekker, New York, 1968.
2. W. Trappe, Biochem. Z., *305*, 150 (1940).
3. M. Steiger and T. Reichstein, Helv. Chim. Acta, *21*, 546 (1938).
4. J. Jacques and J. P. Mathieu, Bull. Soc. Chim. France, *94*, (1946).
5. E. Hecker, Vereilungsnerfrahren in Laboratorium, Supplements to Angermudle Chemig, Verlag Chemie, Weinheim, 1954.
6. R. Neher, in *Thin Layer Chromatography*, G. B. Marini-Bettolo, Ed., Elsevier, Amsterdam, 1964.
7. L. R. Snyder, J. Chromatogr., *63*, 15 (1971).
8. J. J. Kirkland, Ed., *Modern Practice of Liquid Chromatography*, Wiley-Interscience, New York, 1971.
9. P. R. Brown, *High Pressure Liquid Chromatography*, Academic Press, New York, 1973.
10. S. G. Perry, R. Amos, and P. I. Brewer, *Practical Liquid Chromatography*, Plenum Press, New York, 1972.
11. D. L. Saunders, Anal. Chem., *46*, 470 (1974).
12. P. B. Hamilton, Anal. Chem., *35*, 2055 (1963).
13. J. C. Dickinson, H. Rosenbloom, and P. B. Hamilton, Pediatrics, *36*, 2 (1965).
14. S. Mone and W. H. Stein, J. Biol. Chem., *211*, 893 (1954).
15. J. V. Benson, M. J. Gordon, and J. A. Patterson, Anal. Biochem., *18*, 228 (1967).

16.   G. Ertingshausen and H. A. Adler, Amer. J. Clin. Pathol., *53*, 680 (1970).
17.   L. R. Snyder and B. E. Buell, Anal. Chem., *40*, 1295 (1968).
18.   S. Turina and M. Trbojevic, Anal. Chem., *46*, 1988 (1974).
19.   J. C. Giddings, Anal. Chem., *37*, 60 (1965).
20.   D. L. Massart and H. DeClerq, Anal. Chem., *46*, 1988 (1974).
21.   L. R. Snyder and J. J. Kirkland, *Modern Liquid Chromatography*, Wiley-Interscience, New York, 1974.
22.   A. Zaffaroni, R. B. Burton, and E. H. Kentman, J. Biol. Chem., *177*, 109 (1949).
23.   I. M. Hais and L. Macek, *Paper Chromatography*, Academic Press, New York, 1963.
24.   K. J. Hruska and M. Franek, J. Chromatogr., *93*, 475 (1974).
25.   M. H. Blessing and B. Gebele, Res. Exp. Med., *162*, 143 (1974).

CHAPTER 6

Developing Techniques

Development in thin layer chromatography is the process in which mobile phase moves across the sorbent layer to effect separation of the sample substances. It may be accomplished a number of ways with several forms of apparatus, ranging from simple to complex. The most widely used types of development and apparatus will be discussed in this chapter.

6.1 APPARATUS

The majority of thin layer development is done by allowing the solvent (mobile phase) to migrate up (ascend) the plate, which is standing in a suitable sized chamber. Developing chambers may be glass, plastic, or metal; glass chambers are the most common. Depending on plate size, budget, etc., the chamber may be a wide-mouthed screw-cap jar, a battery or museum jar, or a specially constructed chamber or tank for TLC purchased from a general laboratory supply firm or a TLC apparatus company. It can accommodate up to two 20×20 cm plates, or their equivalent, by standing them face to face and generally requires a convenient volume of 100 ml or less of mobile phase.

A tank with a raised glass hump running lengthwise on the inside bottom that acts as a tank divider has been introduced by Camag, Inc. It requires only 20 ml of solution per side and allows two large plates to be developed simultaneously facing each other; this design conserves solvent. The tank has the additional feature of unbreakable stainless steel cover, which is available separately for use on other tanks. The divided bottom allows the tank to be saturated with two systems placed on either side of the divider, and the plate may then be inserted on either side for development. This tank is illustrated in Figure 6.1.

Brinkmann Instruments offers four different sizes of tanks specially designed for multiplate development. One tank will hold five 20×20 cm plates, another holds up to ten 5×20 cm plates, and a third size will accommodate sixteen 5×10 cm plates. The largest tank is constructed of stainless steel and will hold up to 17 analytical layer (250 μ) plates or eight preparative (2000 μ) plates.

Another commonly used developing apparatus, called a sandwich chamber, is shown in Figure 6.2. This particular apparatus has the disadvantage that it may be used only with 20×20 cm plates, but it has several advantages, including small required solvent

Figure 6.1. Class Developing tank for plates up to and including 20×20 cm sizes. Courtesy of A. H. Thomas Co.

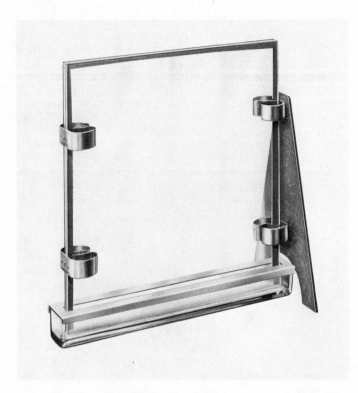

igure 6.2.   Sandwich chamber apparatus for development of 20×20
m plates only.   Courtesy of A. H. Thomas Co.

olume (25 ml), rapid chamber saturation, and fast development.
    The chamber is formed from the 20×20 cm coated thin layer
late to which the sample has been applied, a U-shaped cardboard
pacer, and a clean, uncoated 20×20 cm plate.   The coated plate
ust have an 8-10 mm wide strip of sorbent scraped off on the top
nd sides before the sample is applied.   The cardboard spacer is
laced on this cleared area, and the uncoated, clean plate is
laced on top of the spacer, thereby forming the chamber.   The
ides are clamped together with metal clips, and the open end of
he chamber is placed into a glass solvent trough containing the
eveloping solvent.   Other commercially available sandwich cham-
ers will accommodate smaller (5×20 and 10×20 cm) plates.   These
nambers are easy to make, or they can be purchased commercially
rom a number of suppliers for about the same price as a normal
lass TLC tank.

Analtech, Inc. supplies a "hanging TLC chamber" that allows one or more plates of any size to hang in the chamber containing the mobile phase so that they may equilibrate before being lowered into the solvent system. This may be accomplished without removing the top. The chamber is also suitable for short paper chromatographic strips and sheets.

Brinkmann, Inc. offers the "gas flushing tank" made by Desaga of West Germany shown in Figure 6.3. The intake and effluent stopcocks allow the tank environment to be controlled according to the particular needs of the development process required.

When working with heat-sensitive or unstable compounds, such as many vitamins, it is desirable to work at low temperatures. Thin layer chromatography may be performed at low temperatures by working with the plates and development chamber in a cold room.

Figure 6.3.   Gas flushing tank for flushing of the atmosphere in the tank.   Courtesy of Brinkmann Instruments, Inc.

Brinkmann has a "kryobox" low-temperature developing tank in their
apparatus line, if cold-room facilities are not available. The
developing tank itself is surrounded by a plastic jacket, which
may be insulated, through which a constant-temperature liquid is
passed. It is fitted with a dial-reading thermometer. The developing tank contains a dual plate holder that may be used to equilibrate the plate to the mobile-phase temperature and atmosphere
before it is lowered by external control into the mobile phase.

The Gelman Instrument Company has a self-contained disposable TLC system as part of the Instant Thin Layer Chromatography
(ITLC) line. This system is made up of a plastic Seprachrom mi-
crochromatography chamber, which contains one of the ITLC glass
fiber-impregnated silica gel plates (see Chapter 3). In use, the
top of the Seprachrom chamber is pulled from the base and the sample is applied through a hole in the chamber wall onto the plate
layer. The spots are allowed to dry. Four milliliters of mobile
phase is added to the base, and the chamber is attached, allowed
to equilibrate, and then carefully positioned for development.
From this point, the chromatogram may be treated in the usual manner. Cost of the chamber with plate is about one dollar.

A special developing chamber has been produced by Pharmacia
Fine Chemicals to be used for the development of Sephadex-coated
thin layer plates. This apparatus is pictured in Figure 6.4.

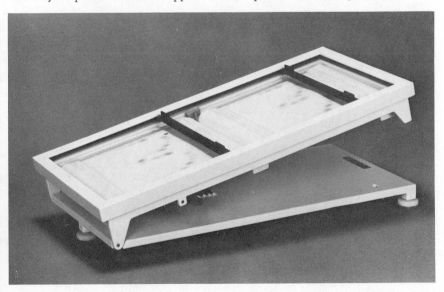

Figure 6.4. Special chamber for the development of Sephadex-
coated plates. Courtesy of Pharmacia Fine Chemicals, Inc.

Special procedures have to be employed for the proper use of
Sephadex-coated plates, and a few of these will be discussed later
in the chapter. The special booklet published by Pharmacia
should be consulted for further details (1).

6.2 DEVELOPMENT MODE

6.2.1 Ascending Development

The majority of thin layer chromatography is done by developing
the plate with the mobile phase, which is moving up (ascending)
the plate. This mode of development is popular because it is the
simplest of the three modes available. The apparatus is readily
available commercially and is easy to use properly. Figures 6.1,
6.2, and 6.3 show apparatus that are used for ascending development. The special chamber pictured in Figure 6.4, which is generally used for the development of Sephadex gel plates, can be
used in the ascending mode as well as in the other two modes,
descending development and horizontal development. Any closed
container that will hold the plate upright is usable.
    There are a number of factors to keep aware of when perform-
ing the development. It has been observed that the angle of development, i.e., the angle at which the plate is supported, affects the rate of development as well as the shape of the spots
(2). As the angle decreases toward the horizontal, the flow of
the mobile phase increases, but spot distortion also increases.
An angle of 75° is recommended as optimum for development.
    If it has been determined that the 75° development angle is
not critical for a particular separation, it is possible to develop a number of plates simultaneously if desired. A tank such
as the one shown in Figure 6.3 may be used. The glass channels
on the inside ends of this tank will support up to seven plates
vertically for development at the same time. If a divided tank
such as this is not available, plate separator supports may be
made by bending a glass rod in the shape of a series of connected U's: ∩∩∩. Each plate can then be supported by placing
its top into a U-section (3). The "hanging TLC chamber" as supplied by Analtech, Inc. may be used to support up to four 20×20
cm plates vertically for simultaneous development.
    If a plate is developed in a chamber that is not saturated
with mobile-phase vapor, unsymmetrical flow up the plate may occur because the mobile phase is evaporating off the plate into
the chamber atmosphere to establish an equilibrium condition.
This entire process depends on the volatility of the mobile phase,
the temperature, and the rate of development. It is for this reason that the development chamber should be as small as convenient

ly possible for the plate size and number of plates being developed at one time. Niederwieser and Honegger (4) note that the
separation process in TLC occurs in a three-phase system of stationary, mobile, and gas phases, all of which are interacting with
each other until an equilibrium is reached. The following qualitative equations sum up what is occurring:

(1) Adsorbent + Solvent   ADSORPTION   Stationary phase +
                                               mobile phase + heat

(2) Mobile phase + air +   EVAPORATION   Gas mixture
     heat

(3) Adsorbent + gas   IMPREGNATION   Impregnated adsorbent
     mixture   + air + heat

Equation 1 describes a very fundamental process for TLC.
While wetting the layer, one portion of the solvent becomes more
highly ordered and together with the layer becomes the stationary
phase. Equations 2 and 3 describe interactions of the mobile
phase that are insignificant with solvents of very low vapor pressure but are important with volatile solvents. Evaporation of
the mobile phase will increase with an increase in the amount of
air or heat in the system, and this will cause increased sample
migration because of greater solvent penetration into the layer.
For consistent, reproducible results it is important to maintain a constant-temperature environment about the development
chamber. Benches that have the sun on them during part of the day
should be avoided because of temperature variation. Development
chambers are of such size that they may be conveniently placed in
a controlled environment space, room or bath in order to maintain
consistency.
It is advisable to weigh down the lid of the chamber so that
the vapor pressure and heat produced by the mobile phase do not
dislodge it and disturb the constant environment and equilibrium
formed within the system.

6.2.2 Chamber Saturation

To aid saturation of the developing chamber atmosphere, the inner
walls are normally lined with blotting paper or a thick chromatography paper such as Whatman 3MM. This is conveniently done by
taking a single large sheet of paper (approximately 33 cm long ×
21 cm high for a normal tank) and placing it into the dry tank so
that the back and ends are lined. The front is left open so the
plate may be observed. If cut large enough to line the three
sides, the paper will stand by itself, and it is best not to use

any glue or tape to support it. The paper should come to within
1 cm of the top of the tank. Be careful that the paper is not so
high that it prevents proper seating of the lid on the tank.
Some separations are best done in an unlined tank.
After carefully measuring the individual components of the
mobile phase to be used, mix them thoroughly in a mixing cylinder,
Erlenmeyer flask, or other container *before* adding them to the developing chamber. Most solvent proportions are measured by volume,
unless stated otherwise in the literature. It is convenient to
pour the mixed mobile phase over the lining paper as you pour it
into the developing chamber. This will speed the saturation of
the paper and the chamber. This is usually done about an hour before the plates are to be developed.
Plate development may be measured in terms of either time or
distance. It may be allowed to proceed for a given period of time,
after which the plate is carefully lifted out of the chamber and
the solvent front marked with a pencil or a sharp object such as a
syringe needle. If the mobile phase is to travel a fixed distance,
for example, 10 cm, this distance is marked on the plate during
the time of sample application. Then when the plate is developing, it is easy to note when the mobile phase has gone the desired
distance and remove the plate from the chamber at this point. A
fixed development distance is used to standardize R_f value, and
because R_f values are between 0.0 and 1.0, 10 cm is a convenient
distance.
Sometimes the solvents present in a mobile phase are not completely miscible with each other. This condition may not be noticed until a second solvent front becomes visible on the plate
during plate development. This occurs because of a process called
solvent demixing, which is more prevalent in adsorption chromatography than in partition chromatography. It results from the preferential retention by the stationary phase of the polar components of the mobile phase (4). A two-component (binary) mobile
phase will produce two zones of fairly constant polarity (polar
near the origin, nonpolar near the solvent front) separated by a
transition zone. If this transition zone is abrupt, it becomes
noticable as a secondary solvent front. Great differences in polarity of the components in a mobile phase are responsible for
such secondary fronts. This occurrence is not necessarily bad,
and the chromatographic process, including visualization, should
be completed before any judgment is made. Often a mobile phase of
this sort is the only system that will provide a given separation,
with the substance in question migrating with the secondary solvent front.
Pittoni and Sussi (5) designed a saturation chamber for TLC
that is smaller in volume than a normal development tank and has
provisions for putting in the mobile phase without removing the

lid. The major drawback to the design, however, is that this
chamber will not accommodate 20×20 cm plates.

DeZeeuw (6,7) and others have studied the influence of chamber saturation on separation. Some of these factors will be mentioned in Chapter 9.

Dhont (8) has recently described an apparatus that allows
TLC plates to reach equilibrium with a well-defined vapor phase
without coming into contact with atmospheric moisture. The apparatus is a nice, easy-to-use design that will accept only 5×20 cm
plates. It may be possible to scale up the apparatus to accept
larger plates.

6.2.3. Descending Development

Descending development is widely employed in paper chromatography
but is very seldom used in TLC. Reasons for this include the complexity of the apparatus, lack of commercial apparatus, and no apparent advantages other than speed of separation for most separations. The same can be stated for horizontal development in TLC,
which is used mostly in Sephadex development. Discussion of these
two development modes is therefore confined to Section 6.6, Sephadex development.

6.3 DEVELOPMENT TECHNIQUES

The simplest development technique in thin layer chromatography is
that of developing a plate once in a pure solvent used as the mobile phase. An example of this would be the separation of a simple
mixture of colored dyes on an alumina plate using pure benzene as
the mobile phase.

If the nature of the sample substance is completely unknown,
it is best to use a simple, single-component mobile phase in the
primary experiments to determine the separation characteristics of
the unknown. If a single-component mobile phase cannot be found
to provide the separation desired, a multicomponent system may be
necessary, or possibly a different type of development using the
single-component system will work. The following sections will
deal with the development options available to achieve the separation desired.

6.3.1 Multiple Development

The technique of multiple development was first applied to thin
layer chromatography in 1955 (9). If a single development of a
plate indicates a partial separation, the plate may be allowed to
dry and reinserted into the developing chamber for redevelopment.

This may be repeated any number of times until a satisfactory separation is achieved. In effect, it increases the effective length of the layer, so as to increase the likelihood that a separation will occur because the molecules have been able to interact over a greater distance and will resolve themselves in the process. As the mobile phase encounters a substance with a low R_f, it will begin to move it along the layer until it reaches another substance with a greater R_f value and begins to move it. As the mobile phase continues to migrate along the layer, this process is repeated a number of times, and the travel length of the substance with the lower R_f keeps increasing until the mobile phase reaches the substance with the higher R_f value. The distance the mobile phase travels after it reaches this substance is continually decreasing.

A number of researchers have discussed the theoretical aspects of multiple development (10-13). Thoma (12) has exhaustively determined the number of developments necessary to separate two substances of known R_f values by various degrees. These are given in Tables 6.1-6.4. The degree of separation desired should first be decided. Using the table for the degree of separation desired, the R_f value of the faster moving substance is located in the left-hand column. To the right of this value, in the same row, is then located the R_f value of the slower moving substance. The number of developments necessary to separate these two substances is found at the head of the column for the slower moving substance.

Longer than normal (20 ×40 cm) plates with precoated layers are available commercially from a number of suppliers. Often these can be used to provide a greater development distance for a difficult separation. Both analytical (250 μ) and preparative (500, 1000, 1500, 2000 μ) layers are available, as listed in Chapter 3. One problem is encountered when it is desired to use the 20×40 cm plates; tanks large enough to hold the plates during development are not widely available from commercial sources. Brinkmann Instruments, Inc. appears to be the only supplier of tanks specially made for the 20×40 cm plates, and here the development is done over the short distance (20 cm) and not the long one. If it is desired to develop over the 40 cm distance, a container of suitable size, such as a large paper chromatography tank, which is available from the general laboratory supply houses, may suffice. Line the tank with a large sheet of heavy filter or chromatography paper, and allow the tank to equilibrate with the mobile phase for an hour before development.

6.3.2 Stepwise Development

When a mixture contains compounds that differ considerably in po-

TABLE 6.1 NUMBER OF SOLVENT PASSES REQUIRED TO SEPARATE TWO SOLUTES 0.1 × LENGTH OF SUPPORT* (FROM J. A. THOMA (12), REPRODUCED WITH PERMISSION OF THE AUTHOR AND ELSEVIER PUBLISHING CO.)

R_f of Faster Moving Solute × 100	R_f of Slower Moving Solute × 100								
	Number of Solvent Passes for Required Separation								
	2	3	4	5	6	7	8	9-14	Impossible
30	23-21								24-29
29	22-20								23-28
28	21-19	22							23-27
27	20-18	21							22-26
26	19-17	20							21-25
25	18-16	19							20-24
24	17-15	18							19-23
23	16-14	17	18						19-22
22	15-13	16	17						18-21
21	14-12	15	16						17-20
20	13-11	15-14							16-19
19	13-10	14							15-18
18	12-9	13			14				15-17
17	11-8	12		13					14-16
16	10-7	11	12						13-15
15	9-6	10	11						12-14
14	8-5	9	10						11-13
13	7-4	8	9			10			11-12
12	6-3	7	8		9				10-11
11	5-2	6	7	8					9-10
10	4-1	6-5		7					8-9
9	3-1	5-4		6					7-8
8	2-1	4-3		5				6	7
7	1	3-2	4				5		6
6		2-1	3			4			5
5		1	2			3			4
4			1		2			3	
3					1			2	
2								1	

* To determine the number of passes for separation, locate the R_f of the faster moving component in the left-hand column. In the same row to the right of this figure locate the R_f of the slower moving component; the number of passes to separate these two components will then be found at the head of the column of the slower moving component.

146 Developing Techniques

TABLE 6.2 NUMBER OF SOLVENT PASSES REQUIRED TO SEPARATE TWO SOLUTES 0.08 × LENGTH OF SUPPORT* (12)

R_f of Slower Moving Solute × 100

R_f of Faster Moving Solute × 100	Number of Solvent Passes for Required Separation								
	2	3	4	5	6	7	8	9-14	Impossible
30	24-23								25-29
29	23-22	24							25-28
28	22-21	23							24-27
27	21-20	22							23-26
26	20-19	21							22-25
25	19-18	20							21-24
24	18-17	19							20-23
23	17-16	18							19-22
22	17-15		18						19-21
21	16-14		17						18-20
20	15-13	16							17-19
19	14-12	15							16-18
18	13-11	14							15-17
17	12-10	13							14-16
16	11-9	12		13					14-15
15	10-8	11		12					13-14
14	9-7	10	11						12-13
13	8-6	9	10						11-12
12	7-5	8	9						10-11
11	6-4	7	8						9-10
10	5-3	6	7				8		9
9	4-2	5	6			7			8
8	3-1	4	5		6				7
7	2-1	4-3			5				6
6	1	3-2		4					5
5		2-1		3				4	
4		1		2				3	
3				1				2	

* See footnote to Table 6.1 for directions.

TABLE 6.3 NUMBER OF SOLVENT PASSES REQUIRED TO SEPARATE TWO SOL-
UTES 0.06 × LENGTH OF SUPPORT* (12)

R_f of Slower Moving Solute × 100

R_f of Faster Moving Solute × 100	Number of Solvent Passes for Required Separation								
	2	3	4	5	6	7	8	9-14	Impossible
30	25	26							27-29
29	24	25							26-28
28	23	24							25-27
27	23-22								24-26
26	22-21								23-25
25	21-20								22-24
24	20-19								21-23
23	19-18								20-22
22	18-17		19						20-21
21	17-16		18						19-20
20	16-15		17						18-19
19	15-14	16							17-18
18	14-13	15							16-17
17	13-12	14							15-16
16	12-11	13							14-15
15	11-10	12							14-13
14	10-9	11				12			13
13	9-8	10			11				12
12	8-7	9		10					11
11	7-6	8		9					10
10	6-5	7	8						9
9	5-4	6	7						8
8	4-3	5	6						7
7	3-2	4	5						6
6	2-1	3	4					5	
5	1	2	3					4	
4		1	2				3		
3			1			2			
2						1			

* See footnote to Table 6.1 for directions.

TABLE 6.4   NUMBER OF SOLVENT PASSES REQUIRED TO SEPARATE TWO SOLUTES 0.04 × LENGTH OF SUPPORT* (12)

R_f of Faster Moving Solute × 100	R_f of Slower Moving Solute × 100								
	Number of Solvent Passes for Required Separation								
	2	3	4	5	6	7	8	9-14	Impossible
30	27								28-29
29	26								27-28
28	25								26-27
27	24								25-26
26	23								24-25
25	22								23-24
24	21								22-23
23	20								21-22
22	19								20-21
21	18		19						20
20	17		18						19
19	16	17							18
18	15	16							17
17	14	15							16
16	13	14							15
15	12	13							14
14	11	12							13
13	10	11							12
12	9	10							11
11	8	9							10
10	7	8						9	
9	6	7					8		
8	5	6			7				
7	4	5			6				
6	3	4			5				
5	2	3		4					
4	1	2		3					
3		1		2					
2				1					

*   See footnote to Table 6.1 for directions.

larity, a single development with one mobile phase may not provide the desired separation. However, if the plate is developed successively with different mobile phases that have different selectivities or strengths, a separation will probably be forthcoming. This procedure was first used by Stahl (14,15).

The mobile phases may be used in a number of ways. The less polar phase may be used first, followed by a more polar phase, or vice versa. More than two developments may be carried out, and the development distance on the plate may be changed with each development.

For example, the first mobile phase used to separate a given sample mixture may be very polar, such as methanol, in order to separate the polar substances. It is allowed to develop part way up the layer before the plate is removed from the chromatography tank and dried. The second mobile phase, a nonpolar solvent such as cyclohexane, is then used to develop the entire length of the plate in an attempt to resolve in the upper portion of the plate those nonpolar substances that were not resolved with the first, polar-phase development. Depending on the separation, these two developments could be carried out in reverse sequence.

6.3.3 Continuous Development

Continuous development is exactly what the name implies: a continuous flow of mobile phase along the length of the plate, with the mobile phase normally allowed to evaporate off as it reaches the end of the plate. This continuous flow development with evaporation was first used by Mottier and Potterat (9). The evaporation was encouraged by raising the lid of the chromatography tank on one edge. Van Den Eijnden (16) has used a chromatography tank that is not as tall (18.5 cm vs. 23 cm) as a normal tank to allow a normal sized TLC plate to stick out through the halves of the split lid. As the mobile phase evaporates off the end of the plate, a continuous flow of phase is maintained up the layer. Although Van Den Eijnden, in the title to his article (16), calls this "a new technique," an identical split-lid procedure was used by Truter some ten years earlier (17). Other workers (18) used a piece of Saran film for their slotted lid.

Brenner and Niederwieser (19) developed a chamber for the continuous development of a thin layer plate in the horizontal position. It is often called the BN chamber. The sample to be separated is first applied to the plate and the apparatus is assembled. The mobile phase is passed to the origin end of the plate by means of a paper wick from a reservoir. The plate is in contact with a cooling block that circulates cold water to prevent vapor condensation in the chamber. The open end of the chamber, through which the evaporation occurs, may be heated to further the evaporation

of high boiling solvents. The apparatus can be used horizontally
for normal or multiple development if the open end is sealed off
to allow an equilibrium to become established. The BN chamber is
supplied commercially by Brinkmann Instruments, Inc.

6.3.4 Two-Dimensional Development

This developing technique is the application of normal, multiple,
or stepwise development in two dimensions. A very versatile method, it has been widely used in paper chromatography and with certain compound areas in TLC. The best-known applications are those
for the separation of clinically important carbohydrates (20-23)
and amino acids (24). Some of the two-dimensional techniques for
amino acids involve an electrophoretic separation in one dimension
and a TLC separation in the second dimension. Kirchner and co-
workers (25) were the first to apply this method to TLC.

The single sample to be separated is spotted in one corner
of a plate and developed in the normal way for a fixed distance.
The plate is removed from the developing chamber and dried. The
plate is turned 90° and placed into a chamber containing a different mobile phase so that it develops in a direction perpendicular to that of the first development. The line of the partially
resolved sample components from the first development forms the
origin for the second development. This procedure is illustrated
in Figure 6.5. The initial sample application should be done
carefully, so that it is not too near to the edge. A distance of
2.5-3 cm is satisfactory; otherwise the line on which the developed spots lie may be below the solvent level when the plate is
turned to be developed the second time. Versatility of the technique lies in the second development with a different mobile phase
and in the opportunity to treat or modify the layer and samples
before the second development. Stahl first demonstrated such
versatility with his separation-reaction-separation (SRS) procedure, in which he carried out a chemical reaction on the plate
before the second development (26). Known standards can be applied to the second origin before the second development to use
for comparison and calculation of R_f values after development and
visualization.

6.4 ADDITIONAL DEVELOPMENT TECHNIQUES

A few other development techniques have been reported in the literature once or twice but are seldom used, primarily because most
of them are impractical in operation. It is easier to substitute
a more practical technique.

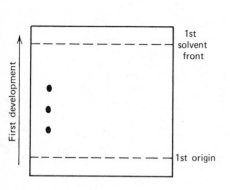

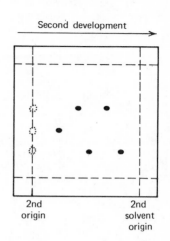

Figure 6.5   Two-dimensional TLC

6.4.1   Centrifugal Chromatography

First applied to paper chromatography, centrifugal chromatography
has also been used in TLC (27,28). Here the TLC plates are made
from circular glass or aluminum with a hole in the center to bolt
onto the centrifuge. The mobile phase is delivered from a reservoir to the plate at a preset rate while the plate spins at 500-
700 rpm. In one instance (27), the development time was decreased
from 35 min to 10 min.

6.4.2   Gradient Elution Development

Gradient elution development is commonly used in column chromatography and occasionally in TLC. In this form of development,
the solvent composition of the mobile phase is changed either in
steps or continuously. Wieland and Determann designed a development chamber for gradient elution TLC (29). It does not appear
to be commercially available. The thin layer plate is supported
on a perforated tray near the bottom of the chamber, which contains a magnetic stirrer bar. The mobile phase is delivered beneath the perforated tray be means of a connecting tube, and an
overflow outlet about a centimeter above the perforated tray keeps
the volume of mobile phase constant. Wieland and Determann used
this apparatus to continuously increase the strength of a buffer
solution (29), as did Hofmann (30). It could be used to change

the solvent composition or pH of the mobile phase or to effect any
other change necessary.

Niederwieser and Honegger have written a review on gradient
techniques in thin layer chromatography (4). It is possible to
use a gradient in the mobile phase, as discussed above, or a gradient in the stationary phase, as Stahl and Dumont have done with
defined pH layers (31,32). They point out that a gradient layer
offers three different directions for mobile phase flow. Chromatograms can be developed with or in the gradient, against the
gradient, or across the gradient. Stahl and Dumont have developed a plate-coating apparatus for the preparation of the gradient
layers. This gradient-mixer (GM) applicator is manufactured by
Desaga and is distributed by Brinkmann Instruments, Inc. The researcher interested in gradient layers should refer to these
papers.

Gradient TLC has been used to separate lipid mixtures having
a wide range of polarity (33,34). Reference 34 describes a simplified gradient procedure requiring only the addition of a glass
solvent trough to a normal-sized TLC developing tank. The paper-
lined developing tank is first saturated with a volatile nonpolar
solvent, referred to as the atmospheric solvent. The trough is
placed in the tank and filled with a relatively polar solvent,
called the trough solvent. The plate to be developed is placed
in the trough. As the trough solvent migrates up the plate and
evaporates, it is slowly replaced by atmospheric solvent. To
produce gradients that are steep enough for lipid analysis, both
evaporation and influx are necessary. The more polar sample components will be retained on the lower portion of the plate as the
polarity of the developing solvent decreases.

Neutral and total lipids have been separated by this procedure. Many combinations of polar trough solvent and nonpolar
atmospheric solvent will produce a suitable gradient for the separation of neutral lipids. Cyclohexane, benzene, and alkanes
with five to eight carbon atoms are good atmospheric solvents.
Suitable trough solvents may be made by mixing two or more polar
solvents or a nonpolar with a polar solvent. The author states
that this method has been successfully combined with reversed-
phase TLC and with argentation TLC. In reversed-phase TLC the
trough contains the less polar solvent. A stationary phase of
50% paraffin oil on kieselguhr with 2:1 butanone-acetonitrile as
the trough solvent and methanol as the atmospheric solvent has
proved successful.

The following practical points may be made:

1. The trough solvent should be volatile enough to evaporate
off the plate but not so volatile that it escapes completely from
the trough.

2. Temperature should be closely controlled.

3. The greater the polarity difference between the trough system and the atmospheric system, the greater the gradient will be.

4. The steepness of the gradient is affected by the rate of development. This depends on the mesh size of the sorbent. Once a suitable sorbent has been found for a given separation, it should be retained.

5. Plate layer thickness should be carefully controlled because of its affect on influx and efflux of the mobile-phase solvents.

6.4.3 Partition Thin Layer Chromatography

Partition TLC is a form of liquid-liquid chromatography in which the solvent coated on the plate is impregnated with the liquid phase before chromatography occurs, and the separation is caused by a partitioning effect between the liquid phase on the sorbent and the mobile phase. The impregnated liquid phase is called the stationary phase. To prevent this phase from changing during development, the mobile phase is usually saturated with the stationary phase.

This is the primary separation process in paper chromatography, where, unless it is intentionally displaced by another liquid, water from the atmosphere is the stationary phase. This process occurs on cellulose thin-layer plates, and many of the development systems and visualization procedures employed in paper chromatography may be directly transposed to cellulose TLC. Cellulose TLC has some advantages compared to paper chromatography, which include ease of handling, smaller space requirements for apparatus, smaller sample size, generally increased detection sensitivity, and faster development. These reasons account for the greater emphasis on TLC compared to paper chromatography.

There are a number of ways of impregnating the sorbent with the stationary phase. Using a hanging TLC chamber, the plate can be equilibrated with the vapor of a mixture of the stationary phase. The stationary phase may also be sprayed on the plate, either by itself or in a volatile solvent as a vehicle. The plate may be dipped in such a solution or allowed to develop in it before the samples are applied. Any volatile solvent is allowed to evaporate before sample application. In normal partition TLC, the stationary phase is a polar substance such as water or formamide and the mobile phase is relatively nonpolar, such as cyclohexane or heptane. Scott has reviewed the theoretical and practical aspects of the stationary phase in TLC (35).

Table 5.3 (page 120) presents a mixotropic solvent series that will serve as a guide for the selection of stationary and mo-

bile phases for partition TLC. See also the appropriate section
in Chapter 5. The polar solvents at the top of the table may be
used as stationary phases and the solvents further down, which are
less polar, as mobile phases. Remember that the closer two solvents are in the series, the more similar are their solubilities
with a given stationary phase and the more similar will be the R_f
values of the separated substances.

6.4.4 Reversed-Phase Partition Development

Reversed-phase development employs a relatively nonpolar stationary phase, which makes the layer hydrophobic, rather than hydrophilic as in normal partition TLC. The phase would be chosen from
the lower portion of Table 5.3. The mobile phase is relatively
polar and may be water or an alcohol. Most commonly used TLC sorbents such as cellulose, silica gel, and kieselguhr have been used
for reversed-phase development. Starch has also been used (36).
This technique is generally useful for the separation of polar
compounds, which often remain at the origin during conventional
development. With reversed-phase development, it is usually the
nonpolar compounds that remain near the origin.

Silica gel treated with dichlorodimethylsilane has been used
for the separation of fatty acids as the free compounds, the
esters, and the hydroxamates (37). Brinkmann Instruments, Inc.
offers plates that are precoated with silanized silica gel 60,
both with and without fluorescent indicator. These plates should
prove useful in the separation of long-chain molecules such as
triglycerides, cholesterol esters, fatty acids and esters, and
carotinoids. For development of the silanized silica gel plates,
it is best to use solvents of medium to high polarity such as
acetonitrile, acetone, ethanol, methanol, acetic acid, formic
acid, and water in various combinations. Higher R_f values will
result as the amount of lipophilic solvent in the eluent is increased. The layer behaves as a macroporous silica gel or as a
deactivated medium porosity silica gel if it is used for adsorption chromatography.

The following points should be observed when using silanized
silica gel plates (38):

1. Eluents (stationary or mobile) containing more than 80%
water should not be used.
2. When the plate is to be impregnated with a stationary
phase containing more than 60% water, it is suggested that the
plate be dried at 120° before the chromatography. This makes the
layer more resistant to water. Prolonged heating should be avoided, to prevent driving the silane groups off the layer.
3. The use of polar eluents results in longer development

times than those normally found with regular silica gel and a less polar eluent.

4. Eluents containing alkaline or ammoniacal components are usable.

5. Eluents containing mineral acid may release minute amounts of polymeric binding agent from a precoated plate. This may interfere with separation or visualization. It is advisable to develop the plate with the acid-containing eluent and dry the plate at 120° for 15 min prior to the chromatography of samples.

6. The fluorescent indicator (F-254) is sensitive to acid, which will reduce the fluorescence. Avoid using acid-containing eluent whenever indicating plates are necessary for visualization.

An apparatus for small-scale reversed-phase TLC that may prove useful has been developed (39).

6.5 AIDS TO DEVELOPMENT

In order to prevent the mobile phase from developing the plate for a greater distance than desired, an electronic sensing device can be used to signal the solvent front at the desired development height. Such a device is available commercially from Applied Science Laboratories and is shown in Figure 6.6. Called the TLC buzzer box, it consists of a probe containing a light and a small photocell that is clipped onto the plate at the desired height of development. When the solvent front reaches the probe, the increased translucency of the sorbent layer causes an increase in the amount of current flowing from the photocell. This current increase activates the warning buzzer in the control box, thus signaling the investigator to remove the plate from the developing chamber. The buzzer can be shut off by pressing the reset button, and plate thickness variations can be compensated for by a simple calibration. Because sorbent translucency is measured, it cannot be used on plates with an opaque support such as aluminum and some plastics.

6.6 DEVELOPMENT OF SEPHADEX PLATES

After the plate is spread with the gel, it is immediately put into the thin layer gel (TLG) chamber for equilibration, *before* the sample is applied. The layer should be connected to the mobile phase reservoirs at both ends by means of filter paper bridges, which should be thoroughly saturated with the eluent before they are connected. The paper bridges should be thick and porous to allow a high flow of solvent. Paper equivalent to Whatman 3MM is most

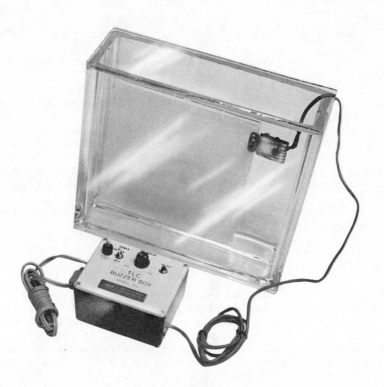

Figure 6.6.   TLC buzzer box for the sensing of the solvent front
on a developing chromatogram.   Courtesy of Applied Science Laboratories, Inc.

suitable.
     Equilibration is carried out with a flow maintained by at
least a 10° angle, for a minimum of 12 hr.   The major function of
equilibration in this process is to normalize the ratio between
the volumes of the stationary and mobile phases.   If the plate has
been prepared with a slurry that is too thin, the layer weight
will decrease during equilibration, indicating a loss of mobile
phase.   If the slurry is too thick, the weight of the layer increases.
     After equilibration the mobile-phase flow is temporarily
stopped to allow the sample to be applied.   The plate must be horizontal during this application, otherwise elongated sample spots
will result.   If the gel layer is disturbed during application,
the flow of the mobile phase will be distorted.
     The concentration of the applied sample is very important in

TLG, as it is in TLC. A high concentration (viscosity) of applied sample will cause zone distortion and influence the migration rate. If different samples have different voscosities, the
migration of the substances relative to each other will vary.
Distortion of each sample zone leads to elongated spots, which
make it difficult to precisely compare and calculate R_f values
and migration distances. Since the flow at the periphery of a
concentrated spot area is faster than at the center, this outer
material will move ahead of the major portion of the spot zone,
producing a leading edge and an elongated spot. It is recommended that protein concentration not exceed 20 mg/ml.

After sample application, flow can be resumed for development of the plate. The flow is maintained by allowing a difference in the levels of the mobile phase between the top and bottom
reservoirs. This is done by tilting the plate or, in the case of
horizontal operation, by adjusting the levels in the two reservoirs. The TLG apparatus is designed so that the plate may be
developed horizontally or at angles between 10 and 45° in 5° intervals. The flow rate is dependent on the angle, the layer
thickness, and the type of gel in the layer.

A flow rate of about 3 cm/hr, obtained at an angle of 10-20°
is generally most suitable. Higher flow rates can cause broaden-
ing of the separated zones in the direction of the development,
with a resulting decrease in resolution. The superfine grades of
Sephadex G-100, G-150, and G-200 will flow between 2.5 and 5 cm/hr
when coated in 0.6 mm layers and set at an angle between 10 and
30°. Superfine grades of G-50 and G-75 will flow 3 cm/hr at even
a 10° angle.

To follow the development process, it is convenient to apply
a marker substance, which may be colored, in a number of positions
across the plate. This substance will show any variation in solvent flow and can also be used for calculation of relative migration distance. Suitable marker substances include myoglobin,
cytochrome c, hemoglobin, Bromphenol blue, labeled albumin, and
Amido Black labeled serum.

6.7 DIAGNOSING A DEVELOPED CHROMATOGRAM

Any of five or six major problems may be encountered if development conditions have not been optimum for a given separation.
Each problem is usually dependent upon three or four major factors. These problems and the factors controlling them will be
discussed here.

Beginners often find that no separation has taken place.
This may be due to the use of an incorrect sorbent, to sample
zones that were not dry before development, to the sample being

lost during preparation, or to the polar component of the mobile phase not being sufficiently fresh.

A distorted solvent front may be due to the movement of the chamber after development has started, or the polar mobile phase component may not be highly enough concentrated or may not be fresh.

A distorted separation, with individual known standards not matching corresponding substances in mixtures, for example, may be due to inadequate mixing of the mobile-phase components before their addition to the development chamber. Other causes include nonalignment of sample zones at the origin when application is being carried out, nonequilibrium in the system, or uneven placement of the plate in the mobile phase.

Undermigration of samples, where R_f values are all low and incomplete or inadequate separation has occurred, may be due to the mobile phase not being polar enough. An inadequate volume of mobile phase, or one that is not reasonably fresh, can also cause this problem. Nonequilibrium may also be a causative factor.

Overmigration of sample zones with resultant high R_f values and inadequate separation often results when the operator does not remove the plate from the chamber after a given development distance. The second most common cause is a mobile phase that is too polar.

REFERENCES

1. "Thin-Layer Gel Filtration with the Pharmacia TLC Apparatus," Pharmacia Fine Chemicals, Inc., Uppsala, Sweden, 1971.
2. D. C. Abbott, H. Egan, E. W. Hammond, and J. Thomson, Analyst, *89*, 480 (1964).
3. M. Brenner and A. Niederwieser, Experientia (Basle), *16*, 378 (1960).
4. A. Niederwieser and C. C. Honegger, in *Advances in Chromatography*, Vol. 2, J. C. Giddings and R. A. Keller, Eds., Marcel Dekker, New York, 1966, p. 125.
5. A. Pittoni and P. L. Sussi, J. Chromatogr., *32*, 422 (1968).
6. R. A. deZeeuw, J. Chromatogr., *32*, 43 (1968).
7. R. A. deZeeuw, J. Chromatogr., *33*, 222 (1968).
8. J. H. Dhont, J. Chromatogr., *90*, 203 (1974).
9. M. Mottier and M. Potterat, Anal. Chim. Acta, *13*, 46 (1955).
10. A. Jeanes, C. S. Wise, and R. J. Dimler, Anal. Chem., *23*, 415 (1951).
11. J. A. Thoma, Anal. Chem., *35*, 214 (1963).
12. J. A. Thoma, J. Chromatogr., *12*, 441 (1963).
13. L. Starka and R. Hampl, J. Chromatogr., *12*, 347 (1963).
14. E. Stahl, Arch. Pharm., *292/64*, 411 (1959).

15. E. Stahl and V. Kaltenbach, J. Chromatogr., *5*, 458 (1961).
16. D. H. Van Den Eijnden, Anal. Biochem., *57*, 321 (1974).
17. E. V. Truter, J. Chromatogr., *14*, 57 (1964).
18. L. M. Libbey and E. A. Day, J. Chromatogr., *14*, 273 (1964).
19. M. Brenner and A. Niederwieser, Experientia, *17*, 237 (1961).
20. R. L. Jolley and M. L. Freeman, Clin. Chem., *14*, 538 (1968).
21. R. L. Jolley and C. D. Scott, Clin. Chem., *16*, 687 (1970).
22. A. S. Saini, J. Chromatogr., *61*, 378 (1971).
23. M. Ghebregzabher, S. Rufini, G. Ciuffini, and M. Lato, J. Chromatogr., *95*, 51 (1974).
24. R. M. Scott, *Clinical Analysis by Thin-Layer Chromatography Techniques*, Ann Arbor-Humphrey, Ann Arbor, 1969, p. 84.
25. J. G. Kirchner, J. M. Miller, and G. J. Keller, Anal. Chem., *23*, 420 (1951).
26. E. Stahl. Arch. Pharmacol., *293/65*, 531 (1960).
27. B. P. Korzun and S. Brody, J. Pharm. Sci., *53*, 454 (1964).
28. J. Rosmus, M. Pavlicek, and Z. Deyl, in *Thin-Layer Chromatography*, G. B. Marini-Bettolo, Ed., Elsevier, Amsterdam, 1964, p. 119.
29. T. Wieland and X. Determann, Experientia, *18*, 431 (1962).
30. A. F. Hofmann, Biochem. Biophys. Acta, *60*, 458 (1962).
31. E. Stahl and E. Dumont, Talanta, *16*, 657 (1969).
32. E. Stahl and E. Dumont, J. Chromatogr. Sci., *7*, 517 (1969).
33. L. R. Snyder and D. L. Saunders, J. Chromatogr., *44*, 1 (1969).
34. G. E. Tarr, J. Chromatogr., *52*, 357 (1970).
35. R. M. Scott, J. Chromatogr. Sci., *11*, 129 (1973).
36. J. Davidek, in *Thin-Layer Chromatography*, G. B. Marini-Bettolo, Ed., Elsevier, Amsterdam, 1964, p. 117.
37. D. Heusser, J. Chromatogr., *33*, 62 (1968).
38. M. Gurkin, personal communication.
39. G. De Vries and V. A. Th. Brinkman, J. Chromatogr., *64*, 374 (1972).

CHAPTER 7

Visualization Procedures

7.1 INTRODUCTION

An ideal visualization or location procedure for thin layer chromatograms should be able to do the following:

1. Visualize microgram quantities of separated substances.
2. Give a visualized area that is firm in its appearance.
3. Give a satisfactory contrast between the visualized area and the background.
4. Give a visualized area that is stable enough and suitable enough for quantitative measurement, if desired.

Very few visualizing agents will do all of these, but it is generally possible to find one that will be suitable for the substances being studied.

7.2 TYPES OF METHODS

Methods for the location (visualization) of substances are of two major types: physical methods, such as the use of ultraviolet (UV) light; and chemical methods, such as reaction with sulfuric acid to produce a brown or black charred area. Two further subclassifications are made within each of these two major types: destructive methods are those that permanently change the chemical identity of the substance being visualized, for example, sulfuric acid charring; and nondestructive methods, which do not produce any permanent change in the chemical identity of the substance be-

visualized. Visualization with UV light is nondestructive with
most substances; ultraviolet light can rearrange some steroid and
vitamin molecules. The use of iodine vapor is an example of a
nondestructive chemical method, and the colored complex formed is
generally nonpermanent and reversible, which leaves the substance
unchanged chemically after the process has occurred. This type of
visualization reaction can be beneficial when it is desired to
follow multi- or two-dimensional development procedures, for instance, or when it is necessary to isolate a substance unchanged
from the plate. The substance can be visualized, its location
marked on the layer physically with a scribe such as a syringe
needle or spatula, the reaction can be allowed to subside, and the
substance may then be scraped from the plate and eluted from the
layer.

7.3   APPARATUS

The majority of visualization procedures may be carried out with
only a few basic pieces of apparatus that are easily afforded by
most laboratories.

For visualization by ultraviolet (UV) light, a number of
long-wave and short-wave lamps are available. Both wavelengths
may be obtained from portable lamps, or from lamps fixed in a
light-tight cabinet, which becomes a convenient, small-scale dark-
room for those laboratories not having such a room to work in.
Figure 7.1 shows such a cabinet in use. For best results, particularly in the submicrogram range, it is necessary to have such
a dark area or dark-room available for ultraviolet visualization.
This cabinet contains both long- and short-wave lamps as well as
white light. The area set aside for such a purpose should be kept
free of dust at all times, and as chemically clean as possible, so
that these forms of contamination may be kept to a minimum.

Visualization reagents are usually placed in contact with the
substances being detected by spraying. Dipping procedures and
impregnation of the plate layer with a visualization reagent before development (2-4) are also used. More will be said about
these later. A number of different spray apparatuses are commercially available. Glass atomizers connected to compressed air
are most generally used, when the air is available. Figure 7.2
shows one such glass atomizer with a rubber bulb attachment,
which can be used when compressed air is not available.

Spray guns using cans of propellant are commercially available. Usually a 4-oz wide-mouthed glass jar is used to contain
the reagent and is screwed onto the spray head. This makes it a
convenient sprayer. The jar used is a standard jar that may be
purchased separately with caps. Such jars may be used to store
the different reagents until they are needed and can then be

Figure 7.1. Ultraviolet light viewing cabinet for the visualization of fluorescent substances on chromatograms. Courtesy of
Brinkmann Instruments, Inc.

attached directly to the spray head without transfer and risk of
contamination. The spray head is thoroughly cleaned after use
with distilled water or other suitable solvent.

Some of the thin layer chromatography suppliers offer a collection of prepared, commonly used reagents in aerosol spray cans.
These are convenient, but often the spray nozzles do not produce
a mist fine enough for uniform coverage.

It is convenient to set aside one section of a fume hood to
be used for the plates. In order to avoid spraying everything in
the hood, a good-sized cardboard box with its top and one side removed may be used to contain the plate while it is being sprayed.
It is best to line this box with heavy filter or chromatography

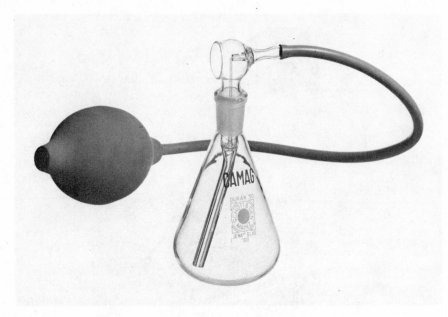

Figure 7.2. Atomizer spray bottle for applying detection reagents
to chromatograms. Courtesy of Camag, Inc.

paper to absorb excess spray and prolong the life of the box. The
plate to be sprayed is simply stood against the back wall of the
box; the open top allows fumes to go up the hood, and the sides
contain the spray within the box. The most inexpensive way to
obtain a box is to locate one being discarded from a store. A
number of TLC suppliers have cardboard or formed plastic spray
boxes available.

When it is preferable for the plate to be dipped into the
visualization reagent, as in the recommended procedure in quantitative thin layer chromatography (5), then it is necessary to
have a container large enough to completely immerse the plate. A
regular TLC tank may be used, but a large volume of reagent is
necessary. More convenient containers that do not have such a
large volume are glass baking dishes and photographic developing
trays of a size suitable to hold the particular size plate being
used. Two different size trays, one for 10×20 cm plates, the
other for 20×20 cm plates, may be kept on hand. It will be necessary to prepare 200-500 ml of reagent to adequately cover the
plate. The preferred technique for dipping is to fill the container with the reagent before the plate is added. Placing a

plate into the container first and then adding the reagent usually produced bad results. The reagent will unevenly penetrate
the layer, producing a nonuniform distribution that often results
in poor visualization and/or quantitation. If the plate layer is
very soft, or the reagent contains too much water, the layer may
bubble or flake off when the reagent is added. Kontes Glass Co.
supplies a small-volume dipping tank for TLC plates. It is shown
in Figure 7.3.

If the plate is impregnated with the visualization reagent
before sample application, the choice of the mobile phase for development must be made very carefully. An incorrect mobile phase
will wash out the reagent during development by carrying it up the
plate with the solvent front. Because of this, the plate is usually dipped after development.

Heating is often necessary to speed the visualization reaction. Most often this can be done at 100-115° for a short period of time such as 10-15 min. Occasionally, higher temperatures and longer periods of time are needed to bring about reaction, such as in some charring techniques. When this is the case,
plate layers containing an inorganic binder such as gypsum, rather
than those made with organic binder, should be used. Organic
binders will char. Since gypsum loses its binding capabilities
at 130°, care must be exercised when heating layers containing
gypsum (such as silica gel G) at temperatures this high.

The plate can be conveniently heated in a small oven placed
near the spraying/dipping location. This oven should be used
only for chemical work and not for delicate operations such as
drying glassware because of the fumes and residues trapped within
the oven. Camag, Inc. offers a hot plate heating apparatus for

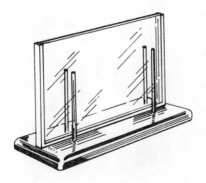

Figure 7.3. Small-volume tank used to dip TLC plates in visualization procedures. Courtesy of Kontes Glass Co.

heating plates uniformly.   It is shown in Figure 7.4.

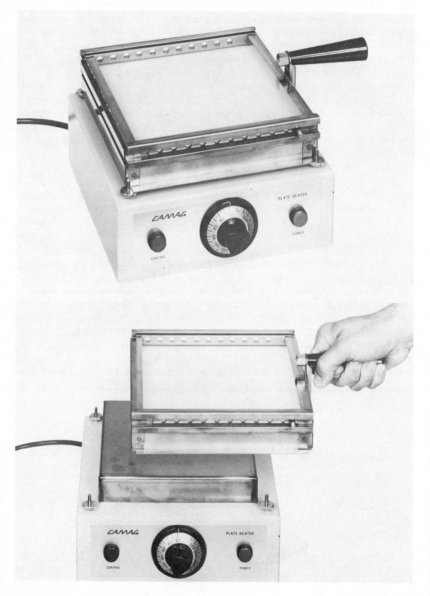

Figure 7.4. Hot plate heating apparatus for promoting reactions on TLC plates.   Courtesy of Camag, Inc.

7.4 REVERSIBLE COLOR REACTIONS

7.4.1 Iodine

It is often advantageous to visualize a substance by a method
that will not permanently alter the substance being visualized.
Such methods would be nondestructive, and many are reversible chemical reactions in which a colored product is formed to allow location of the substance. Often, because of instability, the addition of heat, or other conditions, the visualization reaction
may reverse to the condition prior to treatment and thus allow
subsequent operations to be carried out with the plate. Such necessary operations may include chromatographic development in a
second, or the same, dimension, elution of the substance from the
layer, or *in situ* visualization reaction or quantitation.

Elemental iodine is a simple, rapid, inexpensive, sensitive
and generally nondestructive and reversible visualization reagent
that will stain substances on a chromatogram more readily than the
background. The easiest way to perform iodine visualization is to
place a few crystals of iodine in a covered chromatography tank.
The volatile nature of iodine is such that the tank is readily
filled with purple vapor. The chromatogram to be visualized is
generally dried of any developing solvents and placed into the
iodine-containing tank for a given period of time. Located substances normally turn light brown. The plate is lifted out of the
tank and, if desired, the areas may be physically marked with a
scribe for future reference; otherwise the areas will normally
fade, losing their color completely in a few minutes.

Mangold and Malins (6-8) have readily demonstrated the use of
iodine for the visualization of lipids. The reversible reaction
has also been demonstrated for amino acid derivatives (9), psycho-
active drugs (10), indoles (11), and steroids (12). Many additional compound types have been located with this technique.
Barrett reviewed these (13).

Iodine may react with some compounds. Sensitive ^{131}I techniques (14) and GLC of material extracted from plates (15) have
been used to show the iodination of olefinic double bonds in poly-
unsaturated fatty acids and esters. Lecithin has also been shown
to retain iodine for a considerable time (16), and irreversible
reaction may occur when compounds have a high sensitivity toward
iodine are being detected. Phenolic steroids such as estrone (17)
and drugs such as morphine and oxymorphone (13) apparently form
iodinated derivatives after exposure. In both of these instances,
silica gel plates were being used and it is likely that this cata-
lyzed the iodination (17). Prolonged exposure (18 hr) of alkaloids to iodine vapor produced numberous products (18).

Because of the possibility of undesirable reactions, the

plate should be exposed to the iodine vapor as briefly as is necessary to visualize the substances sought; often only 30 sec will
suffice.

High sensitivity detection is often possible with iodine.
Examples are 0.5 µg estrone (17), 0.1 µg lipid (19), and 0.1 µg
of codeine and morphine (10). In the latter report, the authors
found iodine to be equivalent or superior in the visualization of
these drugs compared to the other three visualization reagents including ammoniacal silver nitrate, potassium permanganate, and
Dragendorff's reagent. Ammoniacal silver nitrate was superior to
iodine in detecting chlordiazepoxide at all levels. Other researchers, investigating sensitivity of lipid detection (19), impregnated silica gel plates with Rhodamine 6G. After developing
and drying the plates, they exposed them to iodine vapor for 2-5
min, followed by UV illumination to reveal intense blue spots.
This combined iodine-UV fluorescence techniques was more sensitive than either method along. Milborrow (20) also found greater
sensitivity with the combined technique.

The iodine reagent may be sprayed on the plates using a dilute (1%) solution of iodine in a volatile, inert solvent.

It is interesting to note that bromine vapor was an unsatis-
factory substitute for iodine in the detection of lipids on TLC
(16). Other researchers (21) working in paper chromatography
noted that bromination occurred, with the formation of hydrogen
bromide from the hydrocarbons being visualized. Bromine was thus
a destructive reagent in this instance.

7.4.2 Water as a Visualization Agent

A number of researchers have used water spray as a nondestructive
visualization agent. The sprayed plate is held against the
light, and hydrophobic compounds are revealed as white areas on a
translucent background. Steroids such as cholestanone and α-
cholestanol were detected in this manner (22). A "heavy spray"
technique consists of saturating the plate with spray and then
allowing it to dry slowly. Sometimes, clearer zones are obtained
in this fashion. This method has been used for cyclohexanols
(23) and hydrocarbons (24). Water visualization has also been
used for bile acids (25,26).

Often, while pulling a wet, developed plate out of a chromatography tank, it will be noted that some separated areas are
visible. As the plate dries, however, these areas are not as
visible. If they are rewetted with solvent, they reappear. Lower limits of detection are often obtained by this method, even
after visualizing a substance by spraying it with a reactive, destructive reagent.

7.4.3 pH Indicators

ommonly used pH indicators such as Bromocresol Green and Brom-
henol Blue are useful for the detection of acidic and basic com-
ounds on chromatograms, through a reversible acid-base equilib-
ium. The polar nature of the indicator should be such that it
an be removed from the sample area after visualization, if this
s desired.

A dilute (0.01-1.0%) aqueous or aqueous alcohol solution of
he indicator is made, and its pH is adjusted to be close to the
ndpoint before use. Bromphenol Blue (0.5% in 0.2% aqueous cit-
ic acid) produces yellow areas on a blue background with aliphat-
c carboxylic acids (27). Methyl red may also be used (28).
romocresol Green (0.3% in 80% aqueous methanol containing 8 drops
0% NaOH per 100 ml solution) produces yellow spots on a green
ackground with aliphatic carboxylic acids (29,30).

7.5 REAGENTS FOR VISUALIZATION

'he preceding section dealt with reagents for visualization that
.re essentially nondestructive to the substances being visualized.
'he present section will deal with the majority of visualization
·eagents used, those that are "destructive" in their action, chem-
.cally changing the substances being visualized.

To review, the commonly used procedure for visualization is
:o spray the chromatogram after it is developed with the detection
·eagent. Other methods include dipping the plate in the reagent
)efore the sample is applied or after the plate is developed. The
lipping procedures are generally recommended when sensitive *in
;itu* quantitation is to be performed (5).

There are a few reagents used generally for the visualization
)f organic compounds. The most widely used is sulfuric acid (31),
vhich sometimes reacts immediately or upon heating to form brown
)r black charred areas. Often a 50% aqueous solution is used,
rather than the concentrated acid. For very unreactive substances,
5% nitric acid can be added to the concentrated acid to increase
its oxidizing capabilities.

Solutions of sulfuric acid-acetic anhydride in ratios of 1:4
(32) and 5:95 (33) have been used. This is the Liebermann-
Burchard reagent. Touchstone et al. (4) have impregnated silica
gel thin layer plates with sulfuric acid (4% v/v) in MeOH, by
dipping prior to sample application and development. These plates
vere used for the *in situ* quantitation of steroids and fatty acids.
Charring is done by heating at 140° for 20 min after development.
Very similar results are obtained by impregnating the gel plates
with ammonium bisulfate (3).

Phosphomolybdic acid, a reagent suitable for many organics, has also been impregnated into silica gel before sample application. These plates were used for the *in situ* quantitation of steroids (2).

In the following pages, reagents and test procedures are listed according to the compound class being detected, with universal reagents and organic nondestructive reagents listed first. The remainder are grouped alphabetically by compound class. These were compiled from many sources (34-38).

Universal Reagents

1. Sulfuric Acid

   Reagent: Concentrated or 50% aqueous.
   Procedure: Spray, heat at 110-120° for a few minutes.
   Results: Brown to black spots for most organics; often the charred products fluoresce.

2. Sulfuric Acid-Acetic Anhydride

   Reagent: 1 part acid; 3 parts anhydride.
   Procedure: Spray, heat at 110-120°.
   Results: Brown to black charred areas.

3. Sulfuric Acid-Sodium Dichromate

   Reagent: Dissolve 3 g sodium dichromate in 20 ml H_2O, dilute with 10 ml conc. H_2SO_4.
   Procedure: Spray, heat at 110°.
   Results: Brown to black charred areas.

4. Sulfuric Acid-Nitric Acid

   Reagent: 1:1 solution.
   Procedure: Spray, heat at 110°.
   Results: Brown to black charred areas.

Organic Acid, Nondestructive Reagents

5. Bromcresol Green

   Reagent: 0.3% Bromcresol green in 1:4 H_2O:methanol, containing 8 drops 30% NaOH/100 ml.
   Procedure: Spray.

Results: Yellow areas for aliphatic carboxylic acids on a
            green background.

5.   Bromcresol Purple

Reagent: 0.04 g Bromcresol purple in 100 ml 50% C_2H_5OH.
            Adjust to pH 10 with alkali.
Procedure: Spray.
Results: Yellow areas on blue background with dicarboxylic
            acids.

7.   Bromphenol Blue

Reagent: 0.5% Bromphenol blue in 0.2% aqueous citric acid.
Procedure: Spray.
Results: Yellow areas for aliphatic carboxylic acids on a
            blue background.

Alcohols

8.   Cericammonium Nitrate

Reagent: 6% cericammonium nitrate in 2N HNO_3.
Procedure: Dry plate 5 min at 105°, cool before spraying.
Results: Polyalcohols produce brown areas on yellow back-
            ground.

9.   2,2-Diphenylpicrylhydrazyl

Reagent: 15 mg reagent in 25 ml $CHCl_3$.
Procedure: Spray. Heat at 110° for 5-10 min.
Results: Alcohols produce yellow spots on purple back-
            ground. Also used for aldehydes and ketones.

10. 4,4'-Methylenebis(N,N-dimethylaniline)

Reagent: Solution (a): 0.25% reagent in acetone.
            Solution (b): 1% ammonium ceric nitrate in 0.2N
                           HNO_3.
            Mix equal portions of each solution before spray-
            ing.
Procedure: Spray with 1:1 solution, heat at 105° for 5 min.
Results: Light blue spots on a blue background.

11. Vanadium Oxinate

Reagent: Dissolve 0.4 g β-hydroxyquinoline in 50 ml of 1:1
            xylene-glacial acetic acid. Heat on a steam bath

                              to 55°. Add 0.2 g ammonium vanadate while stir-
                              ring. Cool and filter. Solution is stable 3
                              days.
        Procedure: Spray.
        Results: Alcohols produce light red spots on a blue-black
                              background.

12. Vanillin-Sulfuric Acid

        Reagent: Dissolve 3 g vanillin in 100 ml abs. C_2H_5OH.
                              Add 0.5 ml H_2SO_4, stirring well.
        Procedure: Spray and heat at 120°.
        Results: Higher alcohols and ketones produce blue-green
                              spots. Also detects bile acids and steroids.

        Aldehydes and Ketones

13. 2,4-Dinitrophenylhydrazine

        Reagent: Dissolve 0.4 g reagent in 100 ml 2N HCl.
        Procedure: Spray.
        Results: Yellow-red spots.

14. 2,2'-Diphenylpicrylhydrazyl

        Reagent: 15 mg reagent in 25 ml $CHCl_3$.
        Procedure: Spray. Heat at 110° for 5-10 min.
        Results: Yellow spots on a purple background.

15. Hydrazine Sulfate

        Reagent: 1% solution of hydrazine sulfate in 1N HCl.
        Procedure: Spray. Observe in daylight and under UV. Heat
                              at 100° and observe under UV.
        Results: Detects aldehydes.

        Alkaloids

16. Bromcresol Green

        Reagent: 0.05% bromcresol green in C_2H_5OH.
        Procedure: Spray. Expose to NH_3 vapor if no reaction, after
                              which the NH_3 is removed.
        Results: Green or blue spots as the blue background fades.
                              May develop at once or within 30 min.

17. Ceric Sulfate-Trichloroacetic Acid

    Reagent: Boil 0.1 g ceric sulfate in 4 ml H_2O containing
                   1 g trichloroacetic acid, adding H_2SO_4 dropwise
                   until the solution clarifies.
    Procedure: Spray and heat at 110°.
    Results: Defects apomorphine, brucine, colchicine, papav-
                   erine, and physostigmine. Also detects organic
                   iodine compounds.

18. Cinnamaldehyde-Acid

    Reagent: Prepare a fresh solution of 1 g cinnamaldehyde
                   in 100 ml CH_3OH.
    Procedure: Spray plate and place in a tank containing a
                   beaker with a 1:1 solution of conc. H_2SO_4:conc.
                   HCl.
    Results: Detects curane alkaloids.

19. Cobalt(II) Thiocyanate

    Reagent: Dissolve 3 g ammonium thiocyanate and 1 g cobalt
                   (II) chloride in 20 ml H_2O.
    Procedure: Spray.
    Results: Alkaloids and amines appear as blue spots on a
                   white to pink background. The colors will fade
                   after 2 hr, but can be revived by spraying with
                   water.

20. Dimethylaminobenzaldehyde

    Reagent: Dissolve 1 g reagent in 30 ml C_2H_5OH, 3 ml conc.
                   HCl, and 180 ml n-butanol.
    Procedure: Spray.
    Results: Ergot alkaloids produce blue spots.

21. Dragendorff Reagent-Munier Modification

    Reagent: Solution (a):   Dissolve 1.7 g bismuth sub-
                                       nitrate and 20 g tartaric acid
                                       in 80 ml H_2O.
                  Solution (b):   Dissolve 16 g KI in 40 ml H_2O.
                  Stock solution: Mix equal parts of solution (a)
                                       and solution (b). Stable 1
                                       month or more.
    Procedure: The spray reagent is prepared by mixing 5 ml of
                   stock solution with a solution of 10 g tartaric

acid in 50 ml H_2O. Stable a week.

Results: For detection of alkaloids and other nitrogen-
             containing compounds, including antihistamines,
             cyclohexylamines, lactams, lipids.

22. Dragendorff Reagent-Munier and Macheboeuf Modification

Reagent: Solution (a):   Dissolve 0.85 g bismuth sub-
                              nitrate in a solution of 10 ml
                              acetic acid and 40 ml H_2O.
             Solution (b):   Dissolve 8 g KI in 20 ml H_2O.
             Stock solution: Mix equal parts of solution (a)
                              with solution (b).
Procedure: The spray reagent is prepared by mixing 1 ml of
             stock solution with 2 ml acetic acid and 10 ml
             H_2O.
Results: For detection of alkaloids and other nitrogen-
             containing compounds.

23. Formic Acid Vapor

Reagent: Formic acid.
Procedure: Expose plate to formic acid vapor for 1 min, ob-
             serve under UV.
Result: Detects quinine and quinidine as fluorescent blue
             spots.

24. Iodine-Potassium Iodide

Reagent: Prepare a 5% solution of I_2 in 10% KI solution.
             A spray solution is made from 2 parts of this
             solution, 3 parts H_2O, and 5 parts 2N acetic acid.
Procedure: Spray.
Results: Detects alkaloids.

25. Iodoplatinate Reagent

Reagent: Solution (a):   5% platinic chloride in water.
             Solution (b):   10% aqueous KI.
             Spray solution: Mix 5 ml of solution (a) with
                              45 ml of solution (b), dilute to
                              100 ml with H_2O. The addition
                              of conc. HCl, 1 part to 10 parts
                              of spray solution, will often
                              increase the sensitivity of the
                              reagent toward certain drugs
                              such as caffeine.

Procedure: Spray.
Results: Most basic drugs give blue or blue-violet spots,
              which turn brownish-yellow. Compound containing
              only primary or secondary amine groups produce
              bluish white spots.

Amines, Amides, and Related Compounds (see reagent 45)

26. Acetoacetylphenol

Reagent: 1% Solution of acetoacetylphenol in n-butanol.
Procedure: Spray and observe under UV light.
Results: Fluorescent areas for adrenaline and similar com-
              pounds.

27. Alizarin

Reagent: 0.1% Alizarin in ethanol.
Procedure: Spray.
Results: Aliphatic amines and amino alcohols produce vio-
              let areas on a faint yellow background.

28. Calcium Nitrate

Reagent: 5% Calcium nitrate in 95% ethanol.
Procedure: Spray and observe under UV.
Results: Diphenylamine produces a yellow-green spot.

29. p-Dimethylaminobenzaldehyde

Reagent: Spray solution (a):   1 g p-Dimethylamino-
                                         benzaldehyde dissolved in
                                         a mixture of 25 ml conc.
                                         HCl and 75 ml CH_3OH.
              Spray solution (b):   1 g p-Dimethylamino-
                                         benzaldehyde dissolved in
                                         100 ml of 96% ethanol.
Procedure: Spray plate with solution (a). Warm plate.
              Spray with solution (b), then place in a tank
              saturated with HCl vapor for 3-5 min or spray
              with 25% HCl.
Results: Detects amines such as tryptamine and related
              citrulline, urea, and tryptophan.

30. Diphenylamine-Palladium Chloride

    Reagent: Solution (a):   1.5% diphenylamine in ethanol.
                   Solution (b):   0.1% palladium chloride in 0.2%
                                  NaCl solution.
    Procedure: Spray with a solution of 5 parts solution (a) to
                 1 part solution (b). Expose moist plate to 240
                 mμ UV.
    Results: Nitrosamines produce blue-violet spots.

31. Ferric Chloride-Potassium Ferricyanide

    Reagent: Solution (a):   0.1 M ferric chloride.
                   Solution (b):   0.1 M potassium ferricyanide.
                   Spray solution: Prepare a 1:1 mixture of solu-
                                   tion (a) to solution (b) im-
                                   mediately before use.
    Procedure: Spray. May have to heat plate at 110°.
    Results: Aromatic amines produce blue spots, which turn
               darker, against a light blue background. Also
               detects phenols and phenolic steroids.

32. Glucose-Phosphoric Acid

    Reagent: Dissolve 2 g glucose in a mixture of 10 ml of 85%
              H_3PO_4 and 40 ml H_2O. Add 30 ml ethanol and 30 ml
              n-butanol to this solution and mix well.
    Procedure: Spray. Heat plate 10 min at 115°.
    Results: Aromatic amines are detected.

33. Malonic Acid

    Reagent: Dissolve 0.2 g malonic acid and 0.1 g sali-
              cylaldehyde in 100 ml absolute ethanol.
    Procedure: Spray plate and heat for 15 min at 120°. Observe
              under UV.
    Results: Amines produce yellow spots.

34. Picric Acid (Jaffe Reagent)

    Reagent: Spray solution (a):   1% picric acid in ethanol.
                   Spray solution (b):   5% KOH in ethanol.
    Procedure: Spray the plate with solution (a) and allow it to
              dry. Then spray with solution (b).
    Results: Substances such as creatinine, glycocyamidine and
               lactams of other α-guanidino acids produce orange
               spots on a yellow background.

35. Potassium Ferrocyanide-Cobalt Chloride

    Reagent: Spray solution (a):   1% aqueous potassium
                                        ferrocyanide.
                 Spray solution (b):   0.5% aqueous cobaltous
                                        chloride.
    Procedure: Spray with solution (a), dry briefly, then spray
                 with solution (b).
    Results: Choline yields a green spot.

36. Potassium Iodate

    Reagent: 1% Aqueous potassium iodate.
    Procedure: Spray, heat at 110° for 2 min.
    Results: Detects phenylethylamines (sympathomimetic
                 amines).

37. Potassium Persulfate-Silver Nitrate

    Reagent: 1% Aqueous potassium persulfate containing
                 1×10^{-3}M silver nitrate.
    Procedure: Spray and heat at 45° if necessary.
    Results: Many compounds will react without heating. Var-
                 ious aromatic amines may be detected; o-toluidine
                 produces blue color changing to yellow; benzidine,
                 brown; p-anisidine, green; p-toluidine, pink; and
                 p-aminophenol, violet.

38. p-Quinone

    Reagent: 0.5 g p-Benzoquinone is dissolved in a mixture of
                 10 ml pyridine and 40 ml n-butanol.
    Procedure: Spray.
    Results: Ethanolamine produces red spots immediately after
                 spraying; choline does not react.

39. Sodium Nitroferricyanide (Sodium Nitroprusside)

    Reagent: Solution (a):   5% sodium nitroferricyanide in
                                  10% acetaldehyde.
                 Solution (b):   1% aqueous sodium carbonate.
                 Spray solution: Mix equal parts of each solu-
                                  tion.
    Procedure: Spray.
    Results: Secondary aliphatic amines, morpholine, and
                 diethanolamine produce blue to violet spots.

40. Sulfanilic Acid (Pauly's Reagent)

    Reagent   Solution (a):   1 g sulfanilic acid in 10 ml
                            conc. HCl plus 90 ml H_2O.
              Solution (b):   5% aqueous sodium nitrite.
              Solution (c):   10% aqueous sodium carbonate.
              Spray solution: Mix one part of (a) with one
                            part (b). Allow to stand for 5
                            min, then add 2 parts solution
                            (c). Allow to cool.
    Procedure: Spray. If plate has been developed in an acidic
              mobile phase, it is sprayed first with solution
              (c) and then dried before spraying with the diazo
              spray.
    Results: Catecholamines, other bases, and hydroxycyclic
              acids produce red-orange or yellow positive re-
              actions. Also detects iodophenols and imidazoles

Amino Acids, Peptides, Proteins, Enzymes

41. Ceric-Arsenite Reagent

    Reagent: Solution (a):   10% ceric sulfate tetrahydrate
                            in cold H_2SO_4. The cloudy so-
                            lution should be refrigerated
                            and then filtered. Store in the
                            cold.
              Solution (b):   Dissolve 5 g sodium arsenite in
                            100 ml cold (0°) 1N H_2SO_4, with
                            stirring. Do not let tempera-
                            ture rise during this step, or
                            arsenious oxide will precipi-
                            tate.
              Spray solution: Mix equal parts of each solution
                            immediately before use.
    Procedure: Place a sheet of filter paper equal in size to
              the TLC plate being visualized onto a clean glass
              plate. Use a small amount of the freshly mixed
              reagent to evenly wet the filter paper. Place
              the layer of the plate upon the wetted paper and
              place a second clean plate on top, pressing down
              firmly. Leave in place for 30 min and visualize
              under UV. Spraying reduces the sensitivity of
              the reagent and exposes the user to the toxic
              reagent.
    Results: Amino acids containing iodine produce white spots

on a yellow background and also fluoresce. Other
organic iodine compounds and iodide react. The
sensitivity is 0.1 µg for triiodothyronine and
thyroxine and 0.01 µg for iodine.

42. Dehydroascorbic Acid

Reagent: Dissolve 0.1 g dehydroascorbic acid in 5 ml of
             H_2O at 60°. Dilute to 100 ml with butanol.
Procedure: Spray and heat at 100° for 5 min.
Results: Hydroxyproline, violet-blue, 25 µg; proline, pale
             yellow, 25 µg; phenylalanine, tyrosine, trypto-
             phan, pale pink, 10 µg; other amino acids, pink
             to red, 1-3 µg.

43. N-Ethylmaleimide

Reagent: Solution (a):   0.05M N-ethylmaleimide in abso-
                              lute isopropanol.
             Solution (b):   1.4% aqueous KOH.
Procedure: Spray plate well with solution (a), allow to dry
             15 min, then spray with solution (b).
Results: Detects amino acids containing -SH groups, as
             pink-red spots, often in amounts as low as 0.1
             µg. Also detects S-acetyl derivatives and thio
             lactones, but not disulfides.

44. Fluorescamine (Fluram, Roche)

Reagent: Dissolve 25 mg Fluram in 100 ml of dimethylsulf-
             oxide or dimethylformamide.
Procedure: Spray, observe under UV while wet.
Results: Amino acids fluoresce white.

45. 1,2,3-Indantrione (Ninhydrin)

Reagent: Solution (a):   0.2% Ninhydrin in n-butanol.
             Solution (b):   10% Aqueous acetic acid.
             Spray solution: Mix 95 parts (a) with 5 parts
                              (b).
Procedure: Spray and heat at 110°.
Results: Amino acids detected as pink-red spots; proline
             is yellow. Detects amines.

46. 1,2,3-Indantrione (Ninhydrin) Cadmium Acetate

Reagent: Dissolve 0.1 g ninhydrin, 0.25 g cadmium acetate,

and 1.0 ml glacial acetic acid in 49 ml methanol.
Procedure: Spray and heat at 120°.
Results: Detects amino acids.

47. Isatin-Zinc Acetate

Reagent: Dissolve 1 g isatin and 1.5 g zinc acetate in
100 ml 95% isopropanol by warming to 80°. Cool,
then add 1 ml acetic acid and store the reagent
in the refrigerator.
Procedure: Spray and heat 30 min at 90°. Good color development is also obtained by not heating, allowing the plate to stand at ambient temperature
for 20 hr.
Results: Detects amino acids.

48. Morin (2',3,4',5,7-Pentahydroxyflavone)

Reagent: 0.005-0.05% morin in methanol.
Procedure: Spray, dry plate at 100° for 2 min, and examine
immediately under UV.
Results: Amino acids produce yellow-green fluorescent or
dark areas on a fluorescent background.

49. Naphthalene Black (stain for Sephadex thin layers)

Reagent: Dissolve 1 g naphthalene black in 100 ml of a
mixture of 10 ml glacial acetic acid, 40 ml H_2O,
and 50 ml methanol.
Procedure: Cover developed Sephadex plate with a sheet of
Whatman 3MM filter paper, being careful to ex-
clude air bubbles. Dry for 30 min at 80-90°. Immerse in naphthalene black solution for 30 min,
then rinse in the acetic acid-methanol-H_2O solution without the dye, to remove excess naphthalene
black.
Results: Detects proteins.

50. 1,3-Naphthoquinone-4-Sulfonic Acid Sodium Salt

Reagent: Dissolve 0.2 g reagent in 100 ml 5% sodium carbonate.
Procedure: Spray after fresh preparation of reagent.
Results: Amino acids produce pink-red spots, which may be
intensified by spraying with a solution of 2 ml
5N NaOH in 98 ml ethanol.

51.   Vanillin-Potassium Hydroxide

Reagent: Solution (a):   2% vanillin in n-propanol.
              Solution (b):   1% KOH in ethanol.
Procedure: Spray with (a) and heat at 110° for 10 min, ob-
              serve under long-wave UV. Spray with (b) and
              heat again.
Results: Detects amino acids and amines. After spraying
              with (a), ornithine will exhibit a bright green-
              yellow and lysine a weak green-yellow fluores-
              cense. After spraying with (b), ornithine will
              possess a salmon color that will fade; proline,
              hydroxyproline, pipecolinic (pipecolic) acid,
              and sarcosine will turn red after a few hours.
              Glycine will produce a green-brown spot, and
              other amino acids will turn faint brown.

Antibiotics

52.   Sodium Azide

Reagent: Solution (a):   0.5% soluble starch.
              Solution (b):   3.5% sodium azide in 0.1N iodine
                              solution.
Procedure: Spray with (a), dry, spray with (b).
Results: Detects penicillin, sensitivity 0.2 µg. Penicil-
              lin and penicilloic acids are also detected by
              starch-iodine-iodide reagents, as are cephalo-
              sporins.

Barbiturates

53.   Diethylamine

Reagent: Dissolve 0.5 g copper sulfate in 100 ml methanol.
              Add 3 ml diethylamine, stirring well.
Procedure: Spray.
Results: Thiobarbituric acids produce green spots.

54.   ʠ-Diphenylcarbazone

Reagent: 0.1% ʠ-diphenylcarbazone in 95% ethanol.
Procedure: Spray.
Results: Barbiturates produce purple spots.

55. Silver Nitrate

   Reagent: Solution (a): 1% aqueous silver nitrate.
                  Solution (b): Reagent 54.
   Procedure: Spray with (a), then with (b).
   Results: Barbiturates produce purple spots.

   Carbohydrates

56. o-Aminodiphenyl

   Reagent: Dissolve 0.3 g o-aminodiphenyl and 5 ml o-phos-
                  phoric acid in 95 ml ethanol.
   Procedure: Spray and heat at 110° for 20 min.
   Results: Brown spots with carbohydrates.

57. Aminohippuric Acid

   Reagent: 0.3% Aminohippuric acid in ethanol. Add 3%
                  phthalic acid if reducing disaccharides are sus-
                  pected to be present.
   Procedure: Spray and heat 8 min at 140°. Observe under UV.
   Results: Pentoses and hexoses produce orange-red spots
                  that are more readily seen under UV.

58. Aminophenol

   Reagent: 0.15 g o- or p-aminophenol in 10 ml 50% H_3PO_4,
                  prepared before use.
   Procedure: Spray and heat at 105°.
   Results: Aldoses and ketoses detected 1-5 µg sensitivity.

59. p-Anisaldehyde

   Reagent: Dissolve 1 ml p-anisaldehyde and 1 ml H_2SO_4 in
                  18 ml ethanol.
   Procedure: Spray and heat at 110°.
   Results: Sugar phenylhydrazones produce green-yellow spots
                  in 3 min. Sugars will produce blue, green,
                  violet spots in 10 min. Also detects digitalis
                  glycosides.

60. p-Anisidine Hydrochloride

   Reagent: 3% p-Anisidine hydrochloride in n-butanol.
   Procedure: Spray and heat at 100° for 2-10 min.

Results: Aldohexoses, green-brown spots; ketohexoses, yellow spots; aldopentoses, green spots; uronic acids, red spots.

51. *p*-Anisidine-Phthalic Acid

Reagent: Dissolve 1.23 g *p*-anisidine and 1.66 g phthalic acid in 100 ml methanol.
Procedure: Spray.
Results: Hexoses, green; pentoses, red-violet, sensitivity 0.5 µg; methylpentoses, yellow-green; uronic acids, brown, sensitivity 0.1-0.2 µg.

52. Anthrone

Reagent: Dissolve 0.3 g anthrone in 10 ml boiling glacial acetic acid. Add 20 ml ethanol, 3 ml H_3PO_4, and 1 ml H_2O.
Procedure: Spray and heat 5 min at 110°.
Results: Ketoses and oligosaccharides produce yellow spots.

53. Benzidine

Reagent: 0.2% benzidine in acetic acid.
Procedure: Spray and heat for 10 min at 100°.
Results: Monoaldoses give brown spots.

54. Bromphenol Blue-Boric Acid

Reagent: Dissolve 40 mg Bromphenol blue (or Bromcresol green) in 10 ml ethanol. Add 0.1 g boric acid and 7.5 ml of 1% sodium tetraborate, and dilute to 100 ml with ethanol.
Procedure: Spray.
Results: Sugar alcohols detected as yellow spots on a blue background.

55. 3,5-Diaminobenzoic Acid

Reagent: Dissolve 1 g 3,5-diaminobenzoic acid dihydrochloride in a mixture of 60 ml H_2O and 25 ml of 80% H_3PO_4.
Procedure: Spray, heat at 110°, and observe under UV.
Results: 2-Deoxy sugars produce a green-yellow fluorescence.

66. 3,5-Dinitrosalicylic Acid

    Reagent: Prepare a 0.5% solution of 3,5-dinitrosalicylic
                    acid in 4% NaOH.
    Procedure: Spray and allow plate to air dry. Heat in oven
                    at 100° for 5 min.
    Results: Reducing sugars will produce brown spots on a
                    yellow background; sensitivity is 1 µg.

67. Diphenylamine

    Reagent: 2.3% Solution of diphenylamine in H_2O saturated
                    with n-butanol.
    Procedure: Spray and allow plate to air dry. Heat in oven
                    at 130° for 20 min.
    Results: Aldoses and ketoses produce blue spots.

68. Hydroxylamine Hydrochloride

    Reagent: Solution (a):   1 g hydroxylamine hydrochloride
                                 in 9 ml H_2O.
                Solution (b):   2 g HaOH in 8 ml H_2O.
                Solution (c):   4 g ferric nitrate in 60 ml H_2O
                                 and 40 ml acetic acid.
    Procedure: Spray with 1:1 mixture of (a) and (b). Dry for
                    10 min at 110°. Spray with mixture of 45 ml of
                    (c) and 6 ml conc. HCl.
    Results: Detects sugar acetates.

69. Malonic Acid-Aniline

    Reagent: Dissolve 1 g malonic acid and 1 ml aniline in 100
                    ml methanol.
    Procedure: Spray, allow plate to air dry, and then heat at
                    110°. Observe under UV.
    Results: Sugars produce yellow, gray, or brown spots
                    fluorescent in UV.

70. Naphthoresorcinol

    Reagent: Dissolve 0.2 g naphthoresorcinol in 100 ml etha-
                    nol plus 10 ml H_3PO_4.
    Procedure: Spray and heat at 110° for 10 min.
    Results: Detects glucuronides as blue spots.

71. Periodic Acid-Benzidine

   Reagent: Solution (a):   2.28% aqueous periodic acid
                                   (stable in cold). Mix 1:19 with
                                   acetone (stable a few hours).
                Solution (b):   Dissolve 184 mg benzidine in 95
                                   ml acetone. Add 4.4 ml H_2O and
                                   0.6 ml acetic acid. This will
                                   turn yellow but is stable indef-
                                   initely.
   Procedure: Spray with (a), allow to stand 4 min, then spray
                with (b).
   Results: Sugars will produce white spots on a blue back-
                ground. Nearly all sugars and sugar alcohols
                react. Sucrose, trehalose, and glycosamines do
                not.

72. Phenol-Sulfuric Acid

   Reagent: Dissolve 3 g phenol and 5 ml H_2SO_4 in 95 ml eth-
                anol.
   Procedure: Spray and heat 10-15 min at 110°. Additional
                heating may intensify spots.
   Results: Carbohydrates produce brown spots.

73. m-Phenylenediamine Hydrochloride

   Reagent: 3.6% m-phenylenediamine hydrochloride in 70%
                ethanol.
   Procedure: Spray and heat briefly at 105°. Observe under
                UV.
   Results: Detects reducing sugars as fluorescent spots.

74. Starch-Iodine

   Reagent: Solution (a):   2% aqueous starch.
                Solution (b):   Dissolve 50 mg I_2 in 100 ml 1%
                                   KI (aqueous).
   Procedure: Spray with (a), and place plate into a moist
                chamber at 45-50° for 1 hr. Dry at room tem-
                perature and spray with (b).
   Results: Amylases detected as white spots on a violet or
                brown background.

75. Vanillin-Perchloric Acid

   Reagent: Solution (a):   1% vanillin in ethanol.

Solution (b):   3% aqueous perchloric acid.
Spray solution: Mix (a) and (b) (1:1).
Procedure: Spray and heat 10 min at 85°.
Results: Deoxy sugars produce various colors.

Carboxylic Acids

See reagents 1-7, 21, 40.

76. Benzidine-Metaperiodate

Reagent: Solution (a):   0.1% aqueous sodium metaperio-
                               date .
             Solution (b):   Dissolve 2.8 g benzidine in 80
                               ml of 96% ethanol. To this add
                               a solution of 70 ml H_2O, 30 ml
                               acetone, and 1.5 ml 1N HCl.
Procedure: Spray with (a); dry partially, then spray moist
             plate with (b).
Results: Detects organic acids, sugars, and sugar alco-
             hols.

77. 2,6-Dichlorophenol-Indophenol

Reagent: 0.1% solution of 2,6-dichlorophenol-indophenol
             in 95% ethanol.
Procedure: Spray. Brief heating may aid color development.
Results: Organic acids produce pink spots on a sky blue
             background.

78. Diphenylcarbazone

Reagent: Solution (a):   0.1-0.2% ᶊ-diphenylcarbazone in
                               95% ethanol.
             Solution (b):   0.05N HNO_3 in ethanol.
Procedure: Spray with (a). Allow to dry. Spray with (b)
             to intensify colors.
Results: Purple spots are produced with additional com-
             pounds of unsaturated acids (e.g., with mercury).
             Produces purple spots with acetoxymercuric-
             methoxy derivatives of unsaturated esters and
             barbiturates.

79. Glucose-Aniline (Schweppe Reagent)

Reagent: Solution (a):   Dissolve 2 g glucose in 2 ml

H_2O.
Solution (b):   Dissolve 2 ml aniline in 2 ml
ethanol.
Spray solution: Mix total solution (a) with
total solution (b) in a 100 ml
volumetric flask. Dilute to
volume with n-butanol.
Procedure: Spray and heat 5-10 min at 125°.
Results: Organic acids produce dark brown spots on a white
background.

80. Methyl Red-Bromthymol Blue

Reagent: Dissolve 0.2 g bromthymol blue and 0.2 g methyl
red in a mixture of 400 ml of 95% ethanol and
100 ml of formaldehyde. Adjust pH to 5.2 with
0.1N NaOH.
Procedure: Spray and place plate in NH_3 vapor.
Results: Organic acids detected as yellow spots on a pink
background changing to red-orange spots on a
green background after exposure to NH_3.

81. o-Phenylenediamine-Trichloroacetic Acid

Reagent: Dissolve 50 mg o-phenylenediamine in 100 ml 10%
aqueous trichloroacetic acid.
Procedure: Spray and heat at 100° for not more than 2 min.
Observe under UV.
Results: Detects α-keto acids.

82. Silver Nitrate-Pyrogallol

Reagent: Solution (a): Dissolve 0.17 g silver nitrate
in 1 ml H_2O. Add 5 ml NH_4OH
and dilute to 200 ml with ethanol.
Solution (b): Dissolve 6.5 mg pyrogallol in
100 ml of ethanol.
Procedure: Spray with (a), then with (b).
Results: Detects organic acids.

Drugs

See also barbiturates section.

83. p-Dimethylaminobenzaldehyde

Reagent: 1% p-Dimethylaminobenzaldehyde in 5% HC1.
Procedure: Spray.
Results: Detects sulfonamides.

84. FPN Reagent

Reagent: Solution (a):   5% ferric chloride.
              Solution (b):   20% perchloric acid.
              Solution (c):   50% nitric acid.
              Spray reagent: Mix (a), (b), and (c) in pro-
                               portions 5:45:50.
Procedure: Spray.
Results: Phenothiazines located as orange, red, or blue
              spots.

85. Furfural-Hydrochloric Acid

Reagent: Solution (a):   10% furfural in ethanol.
              Solution (b):   Conc. HC1.
Procedure: Spray lightly with (a), then (b). Exposure to
              HCL vapor is an alternative to spraying with HCL.
Results: Purple to black areas are produced by carbamates
              with free NH_2 groups. n-Substituted carbamates
              do not react. Ethinamate, meprobamate, and
              styramate react.

86. Mandelin's Reagent

Reagent: 1% ammonium vanadate in conc. H_2SO_4. Shake be-
              fore use.
Procedure: Spray and observe. Heating at 85° for 5 min will
              intensify or change the color of reacting drugs.
Results: Detects antihistamines as various colored spots.

87. Sodium 1,2-Naphthaquinone-4-Sulfonate (NZS Reagent)

Reagent: Solution (a):   0.1N NaOH
              Solution (b):   Saturated solution of sodium
                               1,2-naphthaquinone-4-sulfonate
                               in 1:1 ethanol-H_2O.
Procedure: Spray with (a), then (b).
Results: Thiazide drugs appear as orange spots within 15
              min. Basic drugs with primary amine groups also
              react. Barbiturates do not.

General Reagents

See also reagents 1-4.

88. Antimony Trichloride

 Reagent: Saturated solution of antimony trichloride in
               chloroform.
 Procedure: Spray and heat 10 min at 100°.
 Results: Varied colors produced with various compounds.

89. Sodium Fluorescein

 Reagent: Dissolve 50 mg sodium fluorescein in 100 ml 50%
               methanol.
 Procedure: Spray plate and observe under UV.
 Results: Many aromatic and heterocyclic compounds will
               fluoresce.

90. Fluorescein-bromine

 Reagent: Solution (a):   Reagent 89.
               Solution (b):   Bromine.
 Procedure: After spraying with (a), expose to bromine vapor.
 Results: Ethylenic unsaturated or other types of compounds
               that react with bromine produce yellow spots on a
               pink background.

91. Iodine

 Reagent: Iodine crystals (I_2) or saturated solution of I_2
               in hexane.
 Procedure: Place plate in covered tank containing a few
               crystals of iodine that have vaporized, or spray
               plate with I_2 solution.
 Results: Many compounds absorb iodine reversibly to pro-
               duce yellow to brown spots on a faint yellow
               background. Particularly useful for detection
               of unsaturated fatty acids.

Glycosides

2. Diphenylamine

 Reagent: The spray solution consists of 20 ml 10% diphe-
               ylamine in ethanol, 100 ml HCl, and 80 ml acetic

acid.

Procedure: Spray lightly, cover sprayed plate with another
clean glass plate, and heat 30-40 min at 110°
until positive areas appear.

Results: Glycolipids produce blue spots.

93.   Orcinol (Bials Reagent)

Reagent: Dissolve 0.1 g orcinol in 40.7 ml conc. HCl, add
1 ml 1% $FeCl_3$ solution, and dilute to 50 ml.

Procedure: Spray and heat at 80° for 90 min.

Results: Glycolipids produce violet spots.

94.   Trichloroacetic Acid

Reagent: 25% trichloroacetic acid in $CHCl_3$.

Procedure: Spray and heat 2 min at 100°; observe under UV.

Results: *Strophanthus* glycosides produce a yellow fluorescence with a sensitivity of 0.4 µg.

Heterocyclic Oxygen Compounds

95.   Aluminum Chloride

Reagent: 1% aluminum chloride in ethanol.

Procedure: Spray and view under long-wave UV.

Results: Flavonoids produce yellow fluorescent spots.

96.   *p*-Toluenesulfonic Acid

Reagent: 20% solution of *p*-toluenesulfonic acid in methanol or chloroform.

Procedure: Spray and heat plate at 80° for 10 min. Observe
under long-wave UV.

Results: Flavonoids, indoles, and steroids produce fluorescent spots.

97.   Trichloroacetic Acid

Reagent: 4% trichloroacetic acid in $CHCl_3$.

Procedure: Spray and let plate stand 10-30 min.

Results: Menthofuran produces a pink spot.

Hydrocarbons

See reagents 89, 90.

98. Chlorosulfonic Acid

   Reagent: Prepare a solution of 1:2 chlorosulfonic acid-
               glacial acetic acid.
   Procedure: Spray and heat plate 1 min at 130°.
   Results: Detects olefins and sapogenins.

99. Osmium Tetroxide

   Reagent: Osmium tetroxide crystals.
   Procedure: Expose plate to osmium tetroxide vapors in a
               closed tank. For isolated double bond compounds,
               expose 5-10 min; for conjugated double bonds, ex-
               pose 60 min or more.
   Results: Brown to black spots with double-bond hydrocar-
               bons. May be used for lipids and steroids.

100. Tetracyanoethylene

   Reagent: 10% tetracyanoethylene in benzene.
   Procedure: Spray plates as soon as they are removed from
               development chamber. Heat at 100°.
   Results: Aromatic hydrocarbon show different colors.

   Inorganic Compounds

See reagents 41, 78.

101. Alizarin

   Reagent: Solution (a):   Saturated solution of alizarin
                                in ethanol.
               Solution (b):   25% NH_4OH.
               Solution (c):   Glacial acetic acid.
   Procedure: Spray with (a), then with (b), and then with (c)
               to eliminate background color.
   Results: Many ions produce violet to red spots, including
               Ba, Ca, Mg, Al, Ti, Fe, Zn, Li, Th, Zr, Se, NH_4.

102. Ammonium Molybdate-Stannous Chloride

   Reagent: Solution (a):   1% aqueous ammonium molybdate.
               Solution (b):   1% stannous chloride in 10%
                                HCl.
   Procedure: Spray with (a), dry, and spray with (b).
   Results: Phosphate and phosphite ions produce blue spots.

103. Ammonium Sulfide

  Reagent: Saturated aqueous H_2S made alkaline with NH_4OH.
  Procedure: Spray.
  Results: Cations detected as black: Ag, Hg(I), Hg(III),
        Co, Ni; as brown: Au, Pd, Pt, Pb, Bi, Cu, V, Ti;
        as yellow: Cd, As, Sn; as yellow-orange: Sb.

104. Aurintricarboxylic Acid-Ammonium Salt

  Reagent: 1° solution of aurintricarboxylic acid ammonium
        salt in 1° aqueous ammonium acetate.
  Procedure: Spray and expose plate to NH_3 vapors.
  Results: Detects Al, Cr, and Li.

105. Benzidine

  Reagent: Dissolve 50 mg benzidine base or hydrochloride
        in 10 ml acetic acid, and dilute to 100 ml with
        H_2O filter.
  Procedure: Spray and heat at 85° for 30 min.
  Results: Detects Au(III), Ti(III), chromate, Mn(IV).

106. Bromocresol Purple

  Reagent: Dissolve 40 mg bromocresol purple in 100 ml of
        50% ethanol. Adjust to pH 10 with alkali.
  Procedure: Spray.
  Results: Detects halogen anions except F^-. Also detects
        carboxylic acids (test 6).

107. Brucine

  Reagent: Solution (a): 0.02% brucine in 2N H_2SO_4.
        Solution (b): 2N NaOH.
  Procedure: Spray with (a), warm and note colors, spray with
        (b).
  Results: Detects BrO_3, deep red (a), blood red (b); NO_3,
        red then yellow (a), orange-red (b); ClO_3,
        red-brown (a), blood red (b).

108. Cinchonine-Potassium Iodide

  Reagent: Prepare a fresh solution of 1 g cinchonine in
        100 ml hot H_2O that contains a few drops of
        HNO_3. Cool and add 2 g KI.
  Procedure: Spray.

Results: Detects Bi, orange; Ag, Hg(II), Pb, Sb, V, Tl,
              yellow; Cu, brown; Pt, pink.

109. Copper Sulfate-Mercury Ammonium Sulfate

Reagent: Solution (a):   0.1% cupric sulfate-2N H_2SO_4
                                (9:1).
              Solution (b):   2.7% mercuric chloride and 3%
                                ammonium thiocyanate in H_2O.
Procedure: Spray with (a), then with (b).
Results: Detects: Zn, red to violet; Cu, yellow; Fe(III),
                        red; Au, orange-pink; and Co, blue.

110. Dimethylglyoxime

Reagent: 10% dimethylglyoxime in ethanol containing a
              small amount of NH_4OH.
Procedure: Spray.
Results: Ni ion produces a pink-red spot.

111. Diphenylcarbazide

Reagent: Solution (a):   0.1% s-diphenylcarbazide in
                                96% ethanol.
              Solution (b):   25% NH_4OH.
Procedure: Spray with (a), then with (b).
Results: Heavy metals detected including Ni, blue; Co,
              orange-brown; Ag, Pb, Cu, Sn, Mn, brown; Zn,
              purple.

112. Potassium Thiocyanate

Reagent: 1% aqueous potassium thiocyanate.
Procedure: Spray.
Results: Metal cations detected: Au, orange; Pt, Mo,
              orange-red; Hg(I), black; Bi, U, yellow; V, Co,
              blue; Ni, green; Fe, deep red; Cr, purple; Cu,
              green-black.

113. Pyrocatechol Violet

Reagent: 0.1% pyrocatechol violet in ethanol.
Procedure: Spray with reagent after exposing to UV light
              for 20 min.
Results: Organotin compounds produce dark blue spots on
              a gray-black background.

114. **Silver Nitrate**

Reagent: 0.05N Silver nitrate.
Procedure: Spray, expose to UV light for 10 min. Observe
               immediately, as zones darken with time.
Results: Halogens detected a brown-black areas.

115. **Violuric Acid (Acid Violet)**

Reagent: 1.5% aqueous violuric acid.
Procedure: Spray and heat at 100° for 20 min.
Results: Alkali and alkaline earth metals detected: Li,
               red-violet; Na, violet-red; K, violet; Be,
               yellow-green; Mg, yellow-pink; Ca, orange; Sr,
               red-violet; Ba, light red; Co, green-yellow; Cu,
               yellow-brown.

Lipids, Phospholipids

See reagents 1-3, 88, 91, 92.

116. **Ammonium Molybdate-Perchloric Acid**

Reagent: Solution (a):   3 g ammonium molybdate in 25 ml
                                  H_2O.
            Solution (b):   1N HCl.
            Solution (c):   60% $HClO_4$.
            Spray solution: Mix total solution (a) with 30
                                  ml (b) and 15 ml (c).
Procedure: Spray and heat at 105° for 20 min.
Results: Lipids produce blue-black spots.

117. **Bromthymol Blue**

Reagent: 0.1% bromthymol blue in 10% aqueous ethanol made
               just alkaline with NH_4OH.
Procedure: Spray.
Results: Lipids and phospholopids produce blue-green
               areas. Sensitivity 0.1-1 μg.

118. **α-Cyclodextrin**

Reagent: 1% α-cyclodextrin in 30% ethanol.
Procedure: Spray plate and allow it to dry. Place it in a
               humidity chamber for 1 hr at room temperature,
               then expose it to iodine vapor.

Results: Fatty acids and esters, monoglycerides, and sat-
              urated alcohols remain white. Corresponding un-
              saturated compounds become yellow or brown.

119. Dragendorff's Reagent

Reagent: Solution (a):   17% basic bismuth nitrate in
                               20% aqueous acetic acid.
              Solution (b):   40% aqueous KI.
              Solution (c): H_2O.
              Spray solution: 4:1:14 (a):(b):(c). Store (a)
                               and (b) at 4°. Mix before use.
Procedure: Spray.
Results: Phospholipids containing choline produce orange
              or red-orange spots.

120. Ferric Chloride-Sulfosalicylic Acid

Reagent: Dissolve 0.1 g $FeCl_3 \cdot 6H_2O$ and 7 g sulfosalicylic
              acid in 25 ml H_2O, and dilute to 100 ml with 95%
              ethanol.
Procedure: Spray.
Results: Phosphate groups in lipids and other compounds
              are detected as white fluorescent spots on a
              purple background.

121. Gentian Violet-Bromine

Reagent: 0.1% gentian violet (crystal violet) in metha-
              nol.
Procedure: Spray and place in tank containing bromine vapor.
Results: Lipids produce blue spots on a yellow back-
              ground.

122. Phosphotungstic Acid

Reagent: 10% Phosphotungstic acid in ethanol.
Procedure: Spray and heat at 100° for 5-15 min.
Results: Cholesterol and its esters produce red spots.

123. Rhodamine B

Reagent: Solution (a):   0.05-0.1% Rhodamine B in etha-
                               nol or 0.2% in H_2O.
              Solution (b):   3% hydrogen peroxide.
              Solution (c):   10N KOH.

Procedure: Spray with (a) and observe. Also observe under
             UV. Spraying with (b) may enhance the color.
Results: Higher fatty acids detected. A bright red fluo-
             rescence is produced by many lipids. For gly-
             cerides, spray with (a) and then (c); sensitiv-
             ity may be increased by repeat spraying with (c)
             after a few minutes. Lipids produce purple
             spots on a pink background; glycerides yield
             bright spots on a pink-red to blood-red back-
             ground. Food preservatives produce purple spots
             intensified by spraying with (b).

Nitro and Nitroso Compounds

124. Dimethylaminobenzaldehyde-Tin Chloride-Hydrochloric Acid.

Reagent: Solution (a):   Prepare fresh before use. Mix
                              3 ml 15% stannous chloride sol-
                              ution with 15 ml HCl. Add 180
                              ml H_2O.
             Solution (b):   Dissolve 1 g 4-dimethylamino-
                              benzaldehyde in a mixture of
                              30 ml ethanol, 3 ml HCl and 180
                              ml n-butanol.
Procedure: Spray with (a), air dry, spray with (b).
Results: Nitro compounds produce yellow spots. 3,5-Di-
             nitrobenzoyl derivatives of aliphatic amines are
             also detected.

125. Diphenylamine

Reagent: 1% diphenylamine in 95% ethanol.
Procedure: Spray and expose to 254 nm UV.
Results: Nitrate esters produce yellow-green spots. Ex-
             plosives will produce various colors when a 5%
             solution is used.

126. Malonic Acid Diethylester

Reagent: Solution (a):   10% malonic acid diethylester
                              in ethanol.
             Solution (b):   10% aqueous NaOH.
Procedure: Spray with (a), then with (b). Heat at 95° for
             5 min.
Results: Compound produce red-violet spots. 3,5-Dinitro-
             benzoic acid esters are visible under 254 nm UV

          as dark violet spots. Also visible in daylight.

127. α-Naphthylamine

Reagent: 1% α-naphthylamine in ethanol.
Procedure: Spray.
Results: 3,5-Dinitrobenzoates detected as yellow to or-
                ange spots. If the plate is sprayed with 10%
                KOH in methanol following the naphthylamine,
                these compounds will produce red-brown spots.

Nitrogen Heterocyclic Compounds

[Pyrroles, pyrazole derivatives, imidazoles, indoles, pyridine
derivatives, phenoxazines, quinoline derivatives]

128. Anisidine

Reagent: Solution (a):   1% p-anisidine in ethanol con-
                                    taining 1% HCl.
                Solution (b):   2% amyl nitrite in ethanol.
                Solution (c):   0.4 or 2% aqueous NaOH.
Procedure: Spray with (a), then (b), then (c).
Results: Polyphenols and imidazoles produce red or brown
                spots. Hydroxy indoles also react. The 2% NaOH
                is used when applied after Ehrlich reagent in
                multiple reagent sequences.

129. Boric Acid-Citric Acid

Reagent: Dissolve 0.5 g each of boric acid and citric
                acid in 20 ml methanol.
Procedure: Spray and heat 10 min at 100°. Observe under
                UV.
Results: Quinolines detected; 8-hydroxy quinoline gives
                a yellow-green fluorescence.

130. Dimethyldihydroresorcinol

Reagent: Solution (a):   10% dimethyldihydroresorcinol
                                      in ethanol.
                Solution (b):   5% aqueous ferric chloride.
Procedure: Spray with (a), dry, and spray with (b).
Results: Detects aldehydes and ketones of pyridine bases
                as violet or red-brown spots.

131. 2,4-Dinitrophenylhydrazine

    Reagent: 0.4% 2,4-dinitrophenylhydrazine in 2N HCl.
    Procedure: Spray and heat at 105° for 10 min and observe
                 in visible and UV light.
    Results: Tests for α,β-unsaturated ether lipids (plas-
                 malogens). Aldehyde hydrazones are formed when
                 the plasmalogens are heated in the presence of
                 HCl. These react to produce brown, red, or
                 orange-red spots against a yellow background.

132. Ehrlich Reagent

    Reagent: 10% p-dimethylaminobenzaldehyde in conc. HCl.
                 Spray solution: Mix the 10% solution 1:4 with
                                  acetone before use.
    Procedure: Spray. Color development occurs within 20 min.
    Results: Detects indoles, purple; hydroxyindoles, blue;
                 aromatic amines and ureides, yellow; tyrosine,
                 purple-red. Colors may change and fade with
                 time. These may serve as additional identifica-
                 tion aids.

133. Ferric Chloride-Perchloric Acid.

    Reagent: Solution (a):   0.05M ferric chloride.
                 Solution (b):   5% HClO$_4$.
                 Spray solution: Mix 1:50, (a):(b).
    Procedure: Spray
    Results: Indoles produce red spots that turn blue-yellow
                 after 24 hr.

134. Formaldehyde-Schiff's Reagent

    Reagent: Solution (a):   1% formaldehyde.
                 Solution (b):   Schiff's reagent: 1 g fuchsin,
                                  110 ml 1N HCl, and 5 g sodium
                                  bisulfite are mixed are diluted
                                  to 1 liter with H$_2$O.
    Procedure: Spray with (a), evaporate excess at 110°, then
                 spray with (b).
    Results: Triazines produce red spots.

135. Sodium Nitroferricyanide

    Reagent: Solution (a):   2% sodium nitroferricyanide

                                    (nitroprusside).
                    Solution (b):   5% sodium carbonate.
                    Solution (c):   50% acetic acid.
        Procedure: Spray with (a), then with (b). Dry partially
                    and spray with (c).
        Results: Indoles unsubstituted in positions 1, 2, or 3
                    produce blue spots.

        Peroxides

136.   Ammonium Thiocyanate-Ferrous Sulfate

        Reagent: Solution (a):   0.2 g ammonium thiocyanate in
                                     15 ml acetone.
                    Solution (b):   4% aqueous ferrous sulfate.
                    Spray solution: Add 10 ml of (b) to (a) before
                                     use.
        Procedure: Spray.
        Results: Brownish-red spots from peroxides.

137.   N,N-Dimethyl-p-phenylenediamine Dihydrochloride

        Reagent: Dissolve 1.5 g N,N-dimethyl-p-phenylenediamine
                    dihydrochloride in 128 ml methanol, 25 ml H_2O,
                    and 1 ml acetic acid.
        Procedure: Spray.
        Results: Organic peroxides produce purple-red spots.

138.   Potassium Iodine-Starch

        Reagent: Solution (a):   Mix 10 ml of a 4% KI solution
                                     with 40 ml glacial acetic acid.
                                     Add a small pinch of zinc dust.
                    Solution (b):   Prepare fresh 1% aqueous starch.
        Procedure: Filter solution (a) free of zinc dust immediate-
                    ly before use, then spray the plate. After 5
                    min, spray the plate heavily with (b) until the
                    layer is transparent.
        Results: Blue spots for peroxides due to free iodine re-
                    acting with the starch.

        Pesticides

139.   Brilliant Green

Reagent: 0.5% Brilliant green in acetone.
Procedure: Spray and expose plate immediately to bromine
             vapor.
Results: Organophosphorus and triazine herbicide com-
             pounds produce dark green spots.

140. Bromine-Congo Red

Reagent: Solution (a):   0.4% Congo Red dye in 50%
                              ethanol.
             Solution (b):   10% bromine in carbon tetra-
                              chloride.
Procedure: Expose plate to vapors from solution (b) for 20
             sec. Aerate to get rid of the bromine and
             spray with (a).
Results: Thiophosphate pesticides produce red spots on a
             blue background. Sensitivity 0.5 µg.

141. Cupric Chloride

Reagent: Dissolve 2 g cupric chloride in 11 ml ethanol
             containing 2.5 ml conc. HCl.
Procedure: Spray.
Results: Detects Systox and Meta-Systox.

142. Diphenylamine-Zinc Chloride

Reagent: Dissolve 0.5 g each of diphenylamine and zinc
             chloride in 100 ml acetone.
Procedure: Spray and heat at 200° for 5 min.
Results: Detects chlorinated pesticides.

143. Ferric Chloride-Sulfosalicylic Acid

Reagent: Solution (a):   1% 5-sulfosalicylic acid in
                              80% ethanol.
             Solution (b):   0.1% ferric chloride in 80%
                              ethanol.
Procedure: Expose plate to bromine vapor for 10 min. Spray
             with (b), dry 15 min, then spray with (a).
Results: Thiophosphate pesticides produce white spots on
             a mauve background. Sensitivity 5 µg.

144. Methylumbelliferone

Reagent: Solution (a):   0.5% iodine in ethanol.
             Solution (b):   Dissolve 0.075 g 4-methyl-

umbelliferone in 100 ml 50:50
ethanol-H_2O.   10 ml 0.1N
NH_4OH.

Procedure: Spray with (a), observe positive areas, then
spray with (b) and observe under UV.

Results: Detects organophosphorus pesticide.

45.   Potassium Iodine-Phosphoric Acid

Reagent: Solution (a):   5N KI.
Solution (b):   85% H_3PO_4 (conc.).
Spray solution: Mix 1 part (a) with 15 parts
(b) before use.

Procedure: Spray. Color development may take between ½
and 6 hours.

Results: Rotenone produces a blue spot.

46.   Silver Nitrate-Bromphenol Blue

Reagent: Solution (a):   0.5% silver nitrate in ethanol.
Solution (b):   0.2% bromphenol blue plus 0.15%
silver nitrate in 1:1 ethanol-
ethyl acetate.

Procedure: Spray with (a) and heat at 100° for 5 min.
Spray with (b) and heat at 100° for 10 min.

Results: Chlorinated pesticides produce yellow spots on
a blue background.

47.   Silver Nitrate-Phenoxyethanol

Reagent: Dissolve 0.1 g silver nitrate in 1 ml H_2O and
add 10 ml of 2-phenoxyethanol. Dilute to 200 ml
ml with acetone.

Procedure: Spray and dry in hood for 5 min. Heat at 75°
for 15 min. Expose to UV for a short period of
time and observe.

Results: Chlorinated pesticides produce black spots.
Sensitivity 0.01-0.1 µg.

48. o-Toluidine

Reagent: 0.5% o-toluidine in ethanol.
Procedure: Spray and allow to dry. Observe under 254 nm
UV.
Results: Detects chlorinated pesticides as green spots.
Sensitivity 0.5-1 µg.

149. 1,3,5-Trinitrobenzene

    Reagent: Solution (a):   Dissolve 1 g KOH in 10 ml H_2O and dilute to 100 ml with 95% ethanol.

                Solution (b):   Saturate solution of 1,3,5-trinitrobenzene in acetone.

    Procedure: Spray with (a) and heat at 150° for 5 min. Wash cooled plate with acetone to remove organic residues and spray with (b).

    Results: Organic sulfite pesticides detected as pink to red spots.

    Phenols, including Plant

See reagents 31, 40, 128, 163.

150. Ammonium Vanadate-Anisidine

    Reagent: Solution (a):   Saturated, aqueous ammonium vanadate.

                Solution (b):   Dissolve 0.5 g p-anisidine in 2 ml H_3PO_4, dilute to 100 ml with ethanol and filter.

    Procedure: Spray with (a); while plate is still wet, spray with (b). Heat at 80°.

    Results: Phenols produce various colored spots on a pink background.

151. Benzidine, Tetrazotized

    Reagent: Solution (a):   0.5% benzidine in dilute HCl. Triturate 5 g benzidine with 15 ml of conc. (12N) HCl. Dissolve in 980 ml H_2O. Stable one week.

                Solution (b):   10% sodium nitrite.

                Spray solution: Mix equal volumes of (a) and (b) before use. The solution should be clear and yellow.

    Procedure: Spray

    Results: Phloroglucinol-resorcinol type phenols produce red spots. Sensitivity 2-3 µg.

152. p-Dimethylaminobenzaldehyde-Acetic Anhydride

    Reagent: Solution (a):   5% p-dimethylaminobenzaldehyde in acetic anhydride.
                 Solution (b):   Acetone.
                 Spray solution: Mix 1 part (a) with 4 parts (b).
    Procedure: Spray and heat briefly at 130°.
    Results: Aroyl-glycines produce orange or orange-red spots that fluoresce yellow under UV. Useful for location of citric acid cycle acids.

153. Mercury-Nitric Acid (Millon's Reagent)

    Reagent: Digest 1 part mercury with 2 parts fuming HNO_3, and dilute solution with 2 parts H_2O.
    Procedure: Lightly spray and heat at 35°. Repeat if required.
    Results: Detects phenols, phenol ether glycosides, and aromatic methoxy compounds.

154. p-Nitrosodimethylaniline

    Reagent: 1% p-Nitrosodimethylaniline in 50% ethanol.
    Procedure: Spray.
    Results: Phenols produce spots of various colors.

155. Silver Nitrate, Alkaline

    Reagent: Solution (a):   Saturated aqueous silver nitrate-acetone, 1:20.
                 Solution (b):   0.5% NaOH in 80% ethanol.
    Procedure: Spray with (a), allow to dry, spray with (b).
    Results: Dihydroxy and many polyhydroxy compounds produce gray or gray-brown spots. Simple monohydroxy compounds usually react more slowly. Many dihydroxyphenols react before alkali is applied.

156. Starch-Iodate

    Reagent: Solution (a):   11% starch.
                 Solution (b):   1% potassium iodate.
    Procedure: Spray with (a), then (b); while still wet expose plate to UV for 3-4 min.
    Results: Iodophenolic and iodoamino acid compounds produce blue spots that fade quickly.

157. Ultraviolet Light

Procedure: Examine chromatogram under long-wave (360 nm)
            UV light.
Results: Some phenolic acids fluoresce blue, green, pur-
            ple, or yellow. Identification should not be
            made solely on this basis; specific chemical
            tests must also be made.

Plasticizers

158. p-Nitroaniline

Reagent: Solution (a):   0.5 N KOH in ethanol.
            Solution (b):   1 g p-nitroaniline in 200 ml
                              2N HCl.
            Solution (c):   5% sodium nitrite.
            Spray solution: Add (c) to 10 ml of (b) until
                              solution is colorless.
Procedure: Spray plate with (a) and heat at 60° for 15 min.
            Then spray with solution of (b) and (c).
Results: Detects plasticizers.

159. Resorcinol-Sulfuric Acid

Reagent: Solution (a):   20% resorcinol in ethanol, with
                              a pinch of zinc chloride added.
            Solution (b):   4N H_2SO_4.
            Spray solution: 1:1 mixture (a):(b).
Procedure: Spray and heat at 120° for 10 min. Cool and ex-
            pose to NH_3.
Results: Phthalate ester plasticizers detected. Sensi-
            tivity 20 μg.

160. Thymol-Sulfuric Acid

Reagent: Solution (a):   1% thymol in ethanol.
            Solution (b):   4N H_2SO_4.
Procedure: Spray with (a), heat at 90° for 10 min. Spray
            with (b), and heat at 120° for 10 min.
Results: Plasticizers detected.

Steroids, Sterols, Bile Acids

See reagents 1-4, 12, 21, 31, 70, 88, 91, 94, 99.

161. Allen Test

    Reagent: Prepare a solution of 40 parts conc. H_2SO_4, 9
                  parts ethanol, and 1 part H_2O.
    Procedure: Spray carefully. May have to heat.
    Results: 16-Hydroxy steroids and their acetates produce
                  mauve, rose, purple, or yellow spots.

162. Anisaldehyde-Antimony Trichloride

    Reagent: Mix 1 ml p-anisaldehyde with 100 ml saturated
                  antimony trichloride in chloroform. Add 2 ml
                  conc. H_2SO_4. Keep solution at room temperature
                  in the dark for 1.5 hr.
    Procedure: Take off upper layer of reagent mixture and use
                  it to spray plate. Dry 5 min in dark, and heat
                  at 90° for 3 min. Observe in regular and UV
                  light.
    Results: Detects steroids.

163. Anisaldehyde-Sulfuric Acid

    Reagent: 1 ml conc. H_2SO_4 is added to a solution of 0.5
                  ml anisaldehyde in 50 ml acetic acid. Prepare
                  fresh before use.
    Procedure: Spray and heat at 110° until spots fully de-
                  velop. The pink background may be bleached by
                  exposure to steam.
    Results: Detects steroids, phenols, terpenes, and sugars.

164. Benzoyl Chloride

    Reagent: Solution (a):   Dissolve 20 g zinc chloride in
                                  30 ml glacial acetic acid.
                  Solution (b):   50 g benzoyl chloride in enough
                                  chloroform to make 100 ml solu-
                                  tion.
    Procedure: Spray with (a), heat for 5 min at 90°. Cool and
                  spray with (b). Heat at 90° for 2-3 min. Ob-
                  serve under visible and UV light.
    Results: Detects steroids.

165. Formaldehydogenic Reagent

    Reagent: Solution (a):   15 g ammonium acetate in 85 ml
                                  80% methanol.
                  Spray solution: Before use, add the following

to (a):   1 ml acetylacetone,
0.3 ml glacial acetic acid, and
0.1 ml 50% $HClO_4$.

Procedure: Spray and observe under UV. Fluorescent spots
will develop after 10 min and are most intense
after 60 min.

Results: 20-Hydroxy- and 20-ketonic-21-ols produce green-
yellow fluorescent spots.

166. Glycolic Acid

Reagent: 50% aqueous glycolic acid.
Procedure: Spray and heat at 80° for 45 min. Observe under
254 nm UV.
Results: Detects ketals of α,β-unsaturated steroid
ketones.

167. Hydroxylamine-Ferric Chloride

Reagent: Solution (a):   14% hydroxylamine in methanol.
Solution (b):   3.5N KOH in methanol.
Solution (c):   2% ferric chloride in 10% HCl.
Spray solution: 5 parts (a), 4 parts (b),
shake, filter off KCl

Procedure: Spray plate, allow to dry 10 min, then spray
with (c). If brown $Fe(OH)_3$ appears, let dry and
spray again with (c).

Results: Steroid alcohol esters (other than formates),
steroid acid methylesters, and steroid lactones
produce purple spots.

168. Iodoplatinate Reagent

Reagent: Solution (a):   5% platinum chloride in 1N HCl.
Solution (b):   10% KI.
Spray solution: Mix 5 ml (a), 45 ml (b), and
100 ml H_2O. Store in dark.

Procedure: Spray.
Results: Girard's t-hydrazones produce red to orange
spots. Ketosteroids also detected.

169. Methylene Blue

Reagent: Dissolve 25 mg methylene blue in 100 ml 0.5N
H_2SO_4. Before use, dilute 1 part with an equal
part of acetone.
Procedure: Spray.

Results: Steroid sulfates produce white zones on a blue
               background. Further spraying with 0.01%
               rhodamine 6G in $CHC1_3$ yields red spots on a
               blue background. Generally, sensitivity is in-
               creased.

170. Perchloric Acid

Reagent: 2% Aqueous perchloric acid.
Procedure: Spray and heat 10 min at 150°.
Results: Steroids produce brown spots.

171. Phosphomolybdic Acid

Reagent: 5% phosphomolybdic acid in absolute ethanol.
               Filter. Refrigerate.
Procedure: Spray and heat 10-15 min at 120°.
Results: Detects steroids, lipids, antioxidants as blue
               spots on a yellow background. May also be used
               as a dipping solution.

172. Phosphoric Acid

Reagent: Dilute 1 part H_3PO_4 with 1 part H_2O.
Procedure: Spray until plate appears transparent and heat
               at 120° for 20 min. Observe in visible and UV
               light.
Results: Steroids and bile acids produce various colors
               under visible and UV light.

173. Porter-Silber Reagent (Phenylhydrazine·HC1)

Reagent: Dissolve 43 mg phenylhydrazine hydrochloride in
               a solution of 34 ml ethanol, 41 ml H_2SO_4, and
               25 ml H_2O.
Procedure: Spray.
Results: Steroid-21-aldehydes react immediately with the
               reagent.

174. Resorcylaldehyde (2,4-Dihydroxybenzaldehyde)

Reagent: Solution (a):   0.5% resorcylaldehyde in gla-
                                 cial acetic acid.
               Solution (b):   5% H_2SO_4 in glacial acetic acid.
               Spray solution: Mix equal parts of (a) and (b).
Procedure: Spray and heat at 110° until maximum color
               intensity develops.

Results: 16-Dehydrosteroids and their acetates produce
             blue, mauve, red, or orange spots.

175. Sodium Hydroxide

Reagent: 10% NaOH in 60% aqueous methanol.
Procedure: Spray and heat 10 min at 80°. Observe under 360
             nm UV.
Results: Δ^4-3-Ketosteroids produce yellow fluorescence
             in long-wave UV.

176. Tetrazolium

Reagent: Solution (a):   0.5% blue tetrazolium in meth-
                             anol.
             Solution (b):   6N NaOH in H_2O or methanol.
             Spray solution: Equal parts of (a) and (b) are
                             mixed before use.
Procedure: Spray. Heat may be necessary.
Results: Corticosteroids, 16-hydroxy-17-keto steroids and
             2-hydroxy-3-keto steroids produce blue spots.

177. p-Toluenesulfonic Acid

Reagent: 20% p-Toluenesulfonic acid in $CHCl_3$.
Procedure: Spray and heat at 100° for a few minutes. Ob-
             serve under 360 nm UV.
Results: Steroids, flavonoids, and catechins fluoresce in
             long-wave UV light.

178. 2,3,5-Triphenyltetrazolium Chloride

Reagent: Solution (a):   4% triphenyltetrazolium chlo-
                             ride in methanol.
             Solution (b):   1N NaOH in methanol.
             Spray solution: Mix equal parts of (a) and (b).
Procedure: Spray and heat at 110° for 5-10 min.
Results: Steroids produce red spots. Also detects
             glycosides, reducing sugars, and thio acids.

179. Xanthydrol

Reagent: Dissolve 0.2 g xanthydrol in 90 ml ethanol and
             10 ml conc. HCl.
Procedure: Spray.
Results: Some steroids react. Also detects phenolic
             acids and pyrolles.

180. Zimmermann Reagent (m-Dinitrobenzene)

        Reagent: Dissolve 3 g m-dinitrobenzene in 190 ml methanol
                      Add 10 ml propylene glycol, 5 ml H_2O, and a sol-
                      ution of 2.5 g KOH in 20 ml methanol.
        Procedure: Spray. May have to heat at 110°.
        Results: 3- and 17-oxosteroids are detected.

        Steroid Glycosides

See reagents 59, 88, 94, 184.

181. 1,3-Dinitrobenzene

        Reagent: Solution (a):   10% 1,3-dinitrobenzene in
                                        benzene.
                      Solution (b):   6 g NaOH in 25 ml H_2O. Add 45
                                        ml methanol.
        Procedure: Spray with (a), heat at 60°, then spray with (b).
        Results: Cardiac glycosides produce purple spots, turn-
                      ing to blue, They fade rapidly.

182. Trichloroacetic Acid-Chloramine T

        Reagent: Mix 10 ml of freshly prepared 3% aqueous
                      chloramine T with 40 ml of 25% trichloroacetic
                      acid in 95% ethanol.
        Procedure: Spray and heat 5-10 min at 110°. Observe under
                      UV.
        Results: Digitalis glycosides detected as blue spots
                      under UV.

        Sulfur Compounds

See reagents 25, 53.

183. Benzidine

        Reagent: 50 mg benzidine in 100 ml 1N acetic acid.
        Procedure: Spray and heat at 110°.
        Results: Persulfates produce blue spots.

184. Copper Sulfate

        Reagent: 10% copper sulfate.

Procedure: Spray and heat at 120° for 20 min.
Results: Sulfur-containing glycosides produce brown spots
             on a green background.

185. p-Dimethylaminobenzaldehyde

Reagent: 1% p-dimethylaminobenzaldehyde in 5% HCl.
Procedure: Spray.
Results: Detects sulfonamides.

186. Formaldehyde

Reagent: 40% formaldehyde-H_2O-H_2SO_4 (1:45:55).
Procedure: Spray and heat at 120° for 10 min.
Results: Detects phenothiazines.

187. Iodine-Azide

Reagent: Solution (a):   1.27% I_2 in ethanol.
             Solution (b):   3.25% sodium azide in 1:3 H_2O-
                             ethanol.
             Spray solution: Mix equal volumes of (a) and
                             (b).
Procedure: Spray.
Results: Thiols and disulfides produce white spots on
             brown iodine background, which fades. Spots
             will remain visible in UV even with background
             color gone. Thio-ethers do not react. Detects
             thioureas and penicillins.

188. Isatin

Reagent: 0.4% isatin in conc. H_2SO_4.
Procedure: Spray, observe, and heat at 120°. Observe
             again.
Results: Thiophene derivatives detected.

189. 2-Naphthol-Sodium Nitrite

Reagent: Solution (a):   1% sodium nitrite in 1N HCl.
                             Prepare fresh.
             Solution (b):   0.2% 2-naphthol in 1N KOH.
Procedure: Spray with (a), allow to stand 1 min, then
             spray with (b). Heat at 60°.
Results: Sulfonamides and aromatic amines detected.

190. Periodic Acid-Perchloric Acid

    Reagent: Dissolve 10 g periodic acid and a few mg vana-
                  dium pentoxide in 100 ml 70% $HClO_4$. Mix well.
    Procedure: Spray.
    Results: Thiophosphoric acid esters detected.

191. Potassium Ferricyanide-Ferric Chloride

    Reagent: Solution (a):   1% potassium ferricyanide.
                  Solution (b):   1% ferric chloride.
    Procedure: Spray with (a) and then (b). Observe under UV.
    Results: Detects thiosulfates.

192. Resorcinol-Ammonia

    Reagent: 10% resorcinol.
    Procedure: Spray plate and expose to NH_3 vapor.
    Results: Sulfonic acids detected.

193. Sodium Periodate-Benzidine

    Reagent: Solution (a):   1% sodium metaperiodate.
                  Solution (b):   0.5% benzidine (CAUTION!) in
                                  butanol-acetic acid (4:1).
    Procedure: Spray with (a). Wait 4 min. Then spray with
                  (b).
    Results: Bivalent sulfur compounds produce white spots
                  on a blue background.

Terpenes (Azulenes, Terpene Acids)

194. p-Dimethylaminobenzaldehyde-Phosphoric Acid

    Reagent: Dissolve 1 g p-dimethylaminobenzaldehyde in 5
                  g H_3PO_4 and 50 ml acetic acid.
    Procedure: Spray and heat at 100°.
    Results: Azulenes produce blue spots.

195. Phenol-bromine

    Reagent: 50% phenol in carbon tetrachloride.
    Procedure: Spray and expose to bromine vapor.
    Results: Terpenes detected.

196. Phosphotungstic Acid

    Reagent: 25% phosphotungstic acid in ethanol.
    Procedure: Spray and heat at 115° for 2 min.
    Results: Triterpenoids detected. Sensitivity 3-5 µg.

197. Stannic Chloride

    Reagent: 30% stannic chloride in chloroform.
    Procedure: Spray and heat at 90° for 5-20 min. Observe in
                visible and UV light.
    Results: Detects triterpenes, sterols, steroids, phenols,
                and polyphenols.

198. Vanillin

    Reagent: 1% vanillin in conc. H_2SO_4 or 1% vanillin in 50%
                H_3PO_4.
    Procedure: Spray and heat at 120° for 10-20 min. Observe
                in visible and UV light.
    Results: Terpenes detected.

    Vitamins

See reagent 88.

199. Cacotheline

    Reagent: 2% aqueous cacotheline.
    Procedure: Spray and heat at 100°.
    Results: Vitamin C detected as purple spot.

200. 4-Chloro-1,3-Dinitrobenzene

    Reagent: Solution (a):   1% 4-chloro-1,3-dinitrobenzene
                              in methanol.
                Solution (b):   3N NaOH.
    Procedure: Spray with (a), then with (b).
    Results: Nicotinic acid, nicotinamide, and pyridoxol
                detected.

201. *o*-Dianisidine, Diazotized

    Reagent: Solution (a):   5% sodium carbonate.
                Solution (b):   0.5 g o-dianisidine·2HCl in 60
                              ml H_2O.

Solution (c):   5% sodium nitrite.
Solution (d):   5% urea.
Spray solution: Add 6 ml conc. HCl, 12 ml (c),
                  and 12 ml (d) (after 5 min) to
                  (b). Allow the solution to
                  stand 24 hr before use. It is
                  stable 10 days.

Procedure: Spray heavily with (a) and then lightly with the
             spray solution.
Results: T-Tocopherol and δ-tocopherol produce purple and
             red spots, respectively. Other tocopherols do
             not react.

202. Dipicrylamine

Reagent: Add 1 g dipicrylamine and 0.12 g magnesium
             carbonate to 15 ml H_2O. Heat the mixture 15
             min on a boiling H_2O bath and filter.
             Spray solution: Using 0.2 ml of the above sol-
                               ution, add 50 ml methanol, 49
                               ml H_2O, and 1 ml NH_4OH.

Procedure: Spray.
Results: Detects vitamin B_1.

203. α,α'-Dipyridyl-Ferric Chloride

Reagent: Solution (a):   2% α,α'-dipyridyl in chloro-
                               form.
             Solution (b):   0.5% aqueous ferric chloride.
Procedure: Spray with (a), then spray with (b).
Results: Detects vitamin E, phenols, and other compounds
             with reducing properties.

204. Iodine-Starch

Reagent: 0.001-0.005% I_2, with KI, in 0.4% starch solu-
             tion.
Procedure: Spray.
Results: Ascorbic acid detected as white area on blue
             background.

205. Methoxynitroaniline Sodium Nitrate

Reagent: Solution (a):   Dissolve 0.25 g 4-methoxy-2-
                               nitroaniline in 62 ml glacial
                               acetic acid. Dilute to 125 ml
                               with 10N H_2SO_4.

Solution (b):   0.2% sodium nitrite.
Solution (c):   1:1 solution of (a) to (b).
Solution (d):   2N NaOH.
Procedure: Spray with (c) and then (d).
Results: Vitamin C produces a blue spot on an orange
             background.

206. Potassium Ferricyanide

Reagent: Solution (a):   1% potassium ferricyanide.
             Solution (b):   20% NaOH.
             Spray solution: Mix 1:15:5, (a):H_2O:(b).
Procedure: Spray, allow to dry. Observe under UV.
Results: Vitamin B_1 detected.

207. N,2,6-Trichloro-p-benzoquinoneimine

Reagent: 0.1% Reagent in ethanol.
Procedure: Spray and expose to NH_3.
Results: Vitamin B_6 produces blue spots.

7.6 *In Situ* Chemical Reactions

The generally inert characteristics afforded by the layer sorbents
commonly used make them ideal media for *in situ* chemical reactions
other than those normally done for visualization by spraying and
dipping. Such reactions may include oxidation, reduction, acetylation, esterification, hydrolysis, and others. Miller and
Kirchner (39) were probably the first to describe *in situ* reaction
techniques on TLC plates. The sample to be reacted is applied to
the plate, and the reagent is placed on top. The reaction is allowed to proceed and then developed in the normal manner. Applying known reference compounds to the plate before development will
allow comparison of reaction products and unreacted compounds with
the knowns. Reactions may also be run on a microscale in a small
tube or spot test plate and the mixture applied to the TLC plate
along with known reference compounds. This procedure is often
used by the researcher doing synthesis to monitor the progress of
the reaction. One nice feature is the microscale on which the reactions are carried out.
      There are two broadly classified types of reaction products.
In the first, the product is identifiable via a defined reaction.
One such example is the reduction of cinnamaldehyde by lithium
aluminum hydride to form cinnamyl alcohol (39). The second product type would include those that are complex and unidentifiable
and that may serve as merely "fingerprint methods."

Dallas (40) has summarized the many reactions that have been carried out on plates as shown in Table 7.1. The many different ways in which these reactions may be carried out are summarized in Table 7.2 (40). A few comments about some of these procedures follow.

TABLE 7.1 *IN SITU* REACTIONS ON TLC PLATES

Reaction	Reference
1. Acetylation	41, 42, 43, 44
2. Dehydration by H_2SO_4	39
3. Derivative formation	
(a) Acetate	41, 42, 43, 44
(b) Dinitrobenzoate	39
(c) Dinitrophenyl (of amino acid)	45, 46
(d) Dinitrophenylhydrazone	42, 43, 47
(e) Methyl ester	42
(f) Phenyl isocyanate	39
(g) Semicarbazone	39
4. Diazotization	48
5. Esterification of acids	42
6. Halogenation	43, 49, 50, 51
7. Hydrogenation, catalytic	43
8. Hydrolysis with acid	52, 53, 54, 55, 56
9. Hydrolysis with alkali	39, 43
10. Isomerization	57
11. Methanolysis	54, 55, 58
12. Nitration	43
13. Oxidation with chromic acid	39, 43
14. Oxidation with O_2, O_3	51
15. Oxidation with peroxide, peracid	6, 41
16. Photochemical reaction	57
17. Pyrolysis	60
18. Reduction with $FeSO_4$	50
19. Reduction with $LiAlH_4$	39
20. Reduction with $NaBH_4$	43, 60
21. Reduction with Sn + HCl	48

TABLE 7.2 TECHNIQUES FOR *IN SITU* REACTIONS (40)

Code	Method
Reaction Types	
O	Organized, intended
S	Spontaneous, unintended
D	Defined reaction
F	Fingerprint method
Methods of Reaction Application	
AR	Radiation, heat, light
AG	Gas, vapor
AL	Liquid spray
AM	In mobile phase
AS	In stationary phase
Thin Layer Chromatography	
X1	One-dimensional
X2	Two-dimensional
X2 RSS	Two-dimensional with reaction before first development
X2 SRS	Two-dimensional with reaction between developments
SX RSRS	Two-dimensional with reaction before each development
M	Miscellaneous

7.6.1 Oxidations

The sample to be oxidized may be applied at the origin and covered with a solution of chromic anhydride (saturated) in glacial
acetic acid. Application of 30% hydrogen peroxide and exposure
to UV for 10 min may also work (39).

Malins and Mangold (6) separated and oxidized unsaturated
methyl esters in the presence of saturated compounds by developing the plate in peracetic acid-acetic acid-water (2:15:3). The
oxidizing agent was thus present in the mobile phase (method type,
Table 7.2 O/D/AM/X1).

7.6.2 Reductions

A 10% solution of lithium aluminum hydride in anhydrous ether may be used to effect reduction of a sample applied to the TLC plate. Scotney and Truter (50) have used ferrous salts for the *in situ* reduction of peroxides (method type O/D/AL/X2 SRS) to identify the products of autooxidation of lanostenyl acetate.

Yasuda (61) incorporated a zinc reductor in the layer sorbent to reduce nitro groups.

Kaess and Mathis (60) used sodium borohydride for reduction *in situ*, and nitric acid or chromic acid for exidation in the identification of alkaloids (method type O/D/AL/X1). Other reactions, carried out in capillaries before application to the layer, are also discussed.

Elgamal and Fayez (43), working with pharmacological compounds, have described methods for many *in situ* reactions on both alumina and silica gel, including reductions with sodium borohydride.

7.6.3 Esterification

Working with natural waxes, Holloway and Challen (42) used *in situ* acetylation and esterification with methanol reactions to aid in identification (method type O/D/AL/X1). Bennett and Heftmann (62) esterified steroidal sapogenines *in situ* with trifluoroacetic anhydride. The trifluoroacetic acid formed as a by-product was removed in a hood. Riess (63) employed *in situ* methylation for organophosphorus acids.

7.6.4 Methanolysis

In situ methanolysis of lipids and related compounds has proved of use in this area. Kaufmann et al, (64) have used 12% KOH in methanol for the methanolysis of phospholipids on silica gel for both qualitative and quantitative analysis of mixtures (method types O/D/AL/X1, X2 SRS). Oette and Doss (58) have compared *in situ* and "off-plate" methanolysis of lipids (method type O/D/AL/X1). Viswanathan et al. (54, 55) have developed an interesting procedure involving two-dimensional separation and the hydrolysis of plasmalogens and the methanolysis of acyl phosphatide residues *in situ* (method type O/D/AG + AL/X2 RSRS).

7.6.5 Miscellaneous Reactions

Cyclopropene acids were identified and determined by means of their specific reaction with silver nitrate on the plate (method type O/D/AS/X1) (65).

Dallas (44) described the partial acetylation of polyglycerols *in situ* (method type O/D/AL/X2 SRS).

Stedman (47) formed the dinitrophenylhydrazine (DNPH) derivatives of the bromophenacyl esters of some aromatic acids (method type O/D/AL/X1).

Aromatic nitro compounds have been reduced on silica gel by hydrochloric acid and tin, and the amines formed diazotized (48). Baggiolini and Dewald (56) have used HCl vapor for the partial hydrolysis of p-aminobenzoic esters and sulfonamides on silica gel (method type O/D/AG/X1.

7.7 DERIVATIVE FORMATION PRIOR TO TLC

It is often convenient to prepare derivatives of the substances to be separated before they are applied to the plate. A suitable reaction procedure must be established to produce the desired derivative, and then a suitable mobile phase must be found to separate the derivative.

Procedures for the formation of various derivatives of different compound classes may be found in textbooks of qualitative organic analysis, the chromatography literature, and in the literature of the reagent specialty and chromatography suppliers. These commercial firms offer the chemicals and apparatus necessary to produce many derivatives and offer the procedures to go with them.

A number of examples of derivative formation prior to TLC are presented here to illustrate the applicability of these procedures.

The 2,4-dinitrophenylhydrazones of urinary amino acids (66), urinary ketosteroids (67), and plasma ketosteroids (68) have been prepared from the extracted compounds prior to their chromatographic separation. In the two ketosteroid procedures, thin layer densitometry was performed upon the separated derivatives for quantitative analysis.

For urinary ketosteroids (67) the following procedure is applicable: 5 ml of urine is acidified to pH 1 with 5N HCl, saturated with solid ammonium sulfate, and extracted with 2×20 ml of ethyl acetate. The extract is made neutral with 0.5 ml NH_4OH and evaporated to dryness. The residue is shaken in 40 ml of 1% $HClO_4$ in diethyl ether and incubated for at least 15 hr at 39°. A short boiling should precede the tight stoppering of the tube or flask. After incubation, the ether is washed with 5 ml of 5N NaOH and 3×5 ml H_2O, dried over anhydrous sodium sulfate, and evaporated to dryness.

The derivatization reaction is performed by adding to the residue 0.1 ml of 0.2% (w/v) 2,4-dinitrophenylhydrazine in ethyl acetate. The ethyl acetate is evaporated and 1 ml of 0.03% (w/v) trichloroacetic acid in dry benzene is added and kept for

40 min at 40°. After this period, the solution is diluted with benzene to the original volume of the urine so that aliquots can be easily taken.

The hydrazones may be chromatographed on silica gel G using unidimensional double development in chloroform-carbon tetrachloride (2:1) followed by chloroform:dioxane (94:6). Quantities as low as 1 μg are visible as yellow spots.

Stable yellow derivatives of estrogens may also be formed from Fast Violet Salt B (6-benzoylamino-4-methoxy-5-toluidine, diazonium salt) (69,70). These may be separated by TLC and quantitated if desired. The formation of the derivatives allows the determination of as little as 0.05 μg of steroid.

7.8 ULTRAVIOLET LIGHT

Compounds absorbing ultraviolet light may often be located on thin layer chromatograms if they show differential UV absorption relative to the background, of if they reemit the incident light. Such compounds may be visualized by illuminating the plate with a weak source of monochromatic UV light. A number of UV lamps and lamp cabinets are available commercially; one such cabinet is shown in Figure 7.1. These are essentially the "mineral lights" that come in rock-collecting kits. Two wavelengths are commonly used, 254 nm and 366 nm, as different substances may respond best to only one wavelength.

Many compound types may be visualized with UV light including aromatic amines, 366 nm (71); alkaloids, 254 nm (72); chlorinated pesticides with silver nitrate, 254 nm (73); and polycyclin aromatic hydrocarbons, 366 nm (74-76). See the section on spray reagents for further applications of UV light in conjunction with chemical reaction. The visualization of UV-absorbing compounds is best achieved by a fluorescence-quenching technique, in which a fluorescent substance is incorporated in the sorbent layer before development or the chromatogram is sprayed with a fluorescent agent between development and observation under UV light.

Mangold and Malins (6,8) first used 2',7-dichlorofluorescein as a "nondestructive" spray reagent. The reagent is usually made up as a 0.2% solution in ethanol, and most organic compounds yield yellow-green fluorescent spots on a purple background when viewed under 254 nm UV light. This reagent has been widely used because of its general applicability. Kirchner et al. (31) first used inorganic phosphors, such as zinc silicate and zinc cadmium sulfide, incorporated into the layer sorbents.

As mentioned in Chapters 2 and 3, commercially prepared sorbents and thin layer plates are available with incorporated fluorescent indicators: 254 nm, 366 nm, or both.

Commonly used fluorescent spray solutions include, in addition to the previously noted 2',7-dichlorofluorescein, 0.004-0.04% sodium fluorescein (75,77); 0.05% aqueous Rhodamine B (78-81); 0.005-0.05% Morin in methanol; Rhodamine 6G (82); and fluorescamine (Fluram, Roche) (83-86). The use of some of these is discussed in the spray reagents sections.

Often the intensity of fluorescence on a chromatogram may be increased by exposure to ammonia vapor or by spraying with ammoniacal fluorescein solution (87,88). This occurs as a result of differential absorption or reemission of the incident light. Other workers (84,86,89) have found that the intensity may be increased by spraying with a less volatile organic amine such as triethanolamine (10% in methylene chloride) (84,86), or 50% aqueous morpholine (89).

REFERENCES

1. Z. Gregorowicz and J. Sliwiok, Microchem. J., *16*, 480 (1971).
2. J. C. Touchstone, A. K. Balin, T. Murawec, and M. Kasparow, J. Chromatogr. Sci., *8*, 443 (1970).
3. J. C. Touchstone, T. Murawec, M. Kasparow, and W. Wortmann, J. Chromatogr., *66*, 171 (1972).
4. J. C. Touchstone, T. Murawec, M. Kasparow, and W. Wortmann, J. Chromatogr. Sci., *10*, 490 (1972).
5. J. C. Touchstone and T. Murawec, in *Quantitative Thin Layer Chromatography*, J. C. Touchstone, Ed., Wiley-Interscience, New York, 1973.
6. D. C. Malins and H. K. Mangold, J. Amer. Oil Chemists' Soc., *37*, 576 (1960).
7. H. K. Mangold, J. Amer. Oil Chemists' Soc., *38*, 708 (1961).
8. H. K. Mangold and D. C. Malins, J. Amer. Oil Chemists' Soc., *37*, 383 (1960).
9. G. C. Barrett, Nature, *194*, 1171 (1962).
10. J. H. Copenhauser, D. R. Cronk, and M. J. Carver, Microchem. J., *16*, 472 (1971).
11. O. Hutziger, J. Chromatogr., *40*, 117 (1969).
12. J. S. Matthews, V. A. L. Pereda, and P. A. Aquilera, J. Chromatogr., *9*, 331 (1962).
13. G. C. Barrett, in *Advances In Chromatography*, Vol. 11, J. C. Giddings and R. A. Keller, Eds., Marcel Dekker, New York, 1974, p. 151.
14. M. Z. Nichaman, C. C. Sweeley, N. M. Oldham, and R. E. Olsen, J. Lipid Res., *4*, 484 (1963).
15. E. Vioque and R. T. Holman, J. Amer. Oil Chemists' Soc., *39*, 63 (1962).
16. R. P. A. Sims and J. A. G. Larose, J. Amer. Oil Chemists'

Soc., *39*, 232 (1962).

7. W. Brown and A. B. Turner, J. Chromatogr., *26*, 518 (1967).

8. M. Wilk and U. Brill, Arch. Parm., *301*, 282 (1968).

9. H. E. Vroman and G. L. Baker, J. Chromatogr., *18*, 190 (1965).

0. B. V. Milborrow, J. Chromatogr., *19*, 194 (1965).

1. L. Khin and G. Szasz, J. Chromatogr., *11*, 416 (1963).

2. R. J. Gritter and R. J. Albers, J. Chromatogr., *9*, 392 (1962).

3. R. J. Gritter and R. J. Albers, J. Org. Chem., *29*,. 728 (1964).

4. M. E. Tate and C. T. Bishop, Can. J. Chem., *40*, 1043 (1962).

5. H. Ganshirt, Arch. Pharm., *296*, 73 (1963).

6. F. K. Grutte and H. Gartner, J. Chromatogr., *41*, 132 (1969).

7. H. J. Petrowitz and G. Pastuska, J. Chromatogr., *7*, 128 (1962).

8. A. Lynes, J. Chromatogr., *15*, 108 (1964).

9. D. Braun and H. Geenen, J. Chromatogr., *7*, 56 (1962).

0. S. Miyazaki, Y. Suhara, and T. Kobayashi, J. Chromatogr., *39*, 88 (1969).

1. J. G. Kirchner, J. M. Miller, and G. J. Keller, Anal. Chem., *23*, 420 (1951).

2. L. L. Smith and T. Foell, J. Chromatogr., *9*, 339 (1962).

3. W. L. Anthony and W. T. Beher, J. Chromatogr., *13*, 567 (1964).

4. J. G. Kirchner, *Thin Layer Chromatography*, Wiley-Interscience, New York, 1967, pp. 147-191.

5. "Guide to TLC Visualization Reagents," J. T. Baker Chemical Co., Phillipsburg, N.J.

6. "TLC Visualization Reagents and Chromatographic Solvents," Publication JJ-5, Eastman Kodak Co., Rochester, N.Y., 1973.

7. G. Zweig and J. Sherma, Ed., *Handbook of Chromatography*, Vol. II, CRC Press, Cleveland, 1972, pp. 111-172.

8. K. G. Krebs, D. Heusser, and H. Wimmer, in *Thin Layer Chromatography*, E. Stahl, Ed., Springer-Verlag, New York, 1969, pp. 855-905.

9. J. M. Miller and J. G. Kirchner, Anal. Chem., *25*, 1107 (1953).

0. M. S. J. Dallas, J. Chromatogr., *48*, 193 (1970).

1. G. Mathis and G. Ourisson, J. Chromatogr., *12*, 94 (1963).

2. P. J. Holloway and S. B. Challen, J. Chromatogr., *25*, 336 (1966).

3. M. H. A. Elgamal and M. B. E. Fayez, Anal. Chem., *226*, 408 (1967).

4. M. S. J. Dallas, J. Chromatogr. *48*, 225 (1970).

5. G. Pataki, J. Chromatogr., *16*, 541 (1964).

6. G. Pataki, J. Borko, H. Ch. Curtius, and F. Tancredi, Chromatographia, *1*, 406 (1968).

7. E. D. Stedman, Analyst, *94*, 594 (1969).

8. A. R. Thawley, J. Chromatogr., *38*, 399 (1968).

9. M. Wilk and U. Brill, Arch. Pharm., *301*, 282 (1968).

0. J. Scotney and E. V. Truter, J. Chem. Soc., 1968, 199.

51. M. Wilk, U. Hoppe, W. Taupp, and J. Rochlitz, J. Chromatogr., *27*, 311 (1967).
52. H. H. O. Schmid and H. K. Mangold, Biochim. Biophys. Acta, *125*, 182 (1966).
53. L. A. Horrocks, J. Lipid Res., *9*, 469 (1968).
54. C. V. Viswanathan, F. Phillips, and W. O. Lundberg, J. Chromatogr., *38*, 267 (1968).
55. C. V. Viswanathan, M. Basilio, S. P. Hoevet, and W. O. Lundberg, J. Chromatogr., *34*, 241 (1968).
56. M. Baggiolini and B. Dewald, J. Chromatogr., *30*, 256 (1967).
57. W. Schunack and H. Rochelmeyer, Arch. Pharm., *298*, 572 (1965).
58. K. Oette and M. Doss, J. Chromatogr., *32*, 439 (1968).
59. R. N. Rogers, Anal. Chem., *39*, 730 (1967).
60. A. Kaess and C. Mathis, Ann. Pharm. France, *24*, 753 (1966).
61. S. K. Yasuda, J. Chromatogr., *13*, 78 (1964).
62. R. D. Bennett and E. Heftmann, J. Chromatogr., *9*, 353 (1962).
63. J. Riess, J. Chromatogr., *19*, 527 (1965).
64. H. P. Kaufmann, S. S. Radwan, and A. K. S. Ahmad, Fette Seifen Anstrichmittel, *68*, 261 (1966).
65. A. R. Johnson, K. E. Murray, A. C. Fogerty, B. H. Kennett, J. A. Pearson, and F. S. Shenstone, Lipids, *2*, 316 (1967).
66. K. Figge, Clin. Chim. Acta, *12*, 605 (1965).
67. P. Knapstein and J. C. Touchstone, J. Chromatogr., *37*, 83 (1968).
68. P. Knapstein, L. Treiber, and J. C. Touchstone, Steroids, *11*, 915 (1968).
69. J. C. Touchstone, A. K. Balin, and P. Knapstein, Steroids, *13*, 199 (1969).
70. J. C. Touchstone, T. Murawec, M. Kasparow, and A. K. Balin, J. Chromatogr. Sci., *8*, 81 (1970).
71. M. Gillio-Tos, S. A. Previtera, and A. Vimercati, J. Chromatogr., *13*, 571 (1964).
72. A. H. Der Marderosian, H. V. Pinkley, and M. F. Dobbins, Am. J. Pharm., *140*, 137 (1968).
73. M. F. Dobbins and J. C. Touchstone, in *Quantitative Thin Layer Chromatography*, J. C. Touchstone, Ed., Wiley, New York, 1973, p. 293.
74. E. Sawicki, Chemist-Analyst, *53*, 64 (1964).
75. T. Wieland, G. Luben, and H. Determann, Experientia, *18*, 430 (1962).
76. G. M. Badger, J. K. Donnelly, and T. M. Spotswood, J. Chromatogr., *10*, 397 (1963).
77. T. H. Simpson and R. S. Wright, Anal. Biochem., *5*, 313 (1963)
78. C. Michalec, M. Sulc, and J. Mestan, Nature, *193*, 63 (1962).
79. J. Battailie, R. L. Dunning, and W. D. Looms, Biochem. Biophys. Acta, *51*, 538 (1961).
80. J. W. Copius-Peereboom and H. W. Beekes, J. Chromatogr., *14*,

417 (1964).
81. S. Hunek, J. Chromatogr., *7*, 561 (1962).
82. C. B. Scrignar, J. Chromatogr., *14*, 189 (1964).
83. S. Udenfriend, S. Stein, P. Bohlen, W. Dairman, and M. Weigle, Science, *178*, 871 (1972).
84. A. M. Felix and M. H. Jimenez, J. Chromatogr., *89*, 361 (1974).
85. J. Sherma and J. C. Touchstone, Anal. Lett., *7*, 279 (1974).
86. K. Imai, P. Bohlen, S. Stein, and S. Udenfriend, Arch. Biochem. Biophys., *161*, 161 (1974).
87. M. Brenner and A. Niederwieser, Experientia, *16*, 378 (1960).
88. R. C. Hignett, J. Chromatogr., *31*, 571 (1967).
89. L. D. Hunter, J. Chromatogr., *20*, 595 (1965).

CHAPTER 8

Documenting the Chromatogram

8.1 EVALUATION OF THE CHROMATOGRAM

The process of documenting a developed chromatogram consists of two major steps: (a) evaluating the chromatogram and (b) recording the chromatogram. Good and complete evaluation aids recording.

When the developed chromatogram is removed from the developing chamber, it is first observed in visible light and then in short-wave and long-wave ultraviolet light. If the sorbent layer is one of the harder types, such as cellulose, notations may be made with a pencil right on the layer, if desired, after each observation. If the layer is one of the softer types, notations may be conveniently made with a dissecting or disposable syringe needle. The solvent front, and any irregularities in it, should be marked immediately as the plate is being removed from the developing chamber.

When any visible or UV-visible areas have been noted, the chromatogram may be further visualized by chemical reaction by spraying or dipping the plate (if feasible) with suitable reagents as determined from the literature. Any positive areas that develop immediately are noted, and the plate is heated if desired. Once again, any positive areas are noted, including location, color, intensity, etc. The notation of color is often difficult because color assessment is subjective, and shading and intensity have an influence. If this is found to be a problem, a standardized color chart, as used in the paint industry or in philately, may partially alleviate the problem. Nybom (1) devised a color key system for marking fluorescent areas obtained from plant com-

pounds. His system is reproduced here, as it may prove of practical usefulness to other workers. See Figure 8.1.

After noting spot locations and colors, R_f values or hR_f
values ($R_f \times 100$) may be measured. Spots that identify with known
standard substances may be identified by name. Areas that are not
immediately identified may be characterized by R_f values, color
reactions, chemical tests, etc.

8.2 RECORDING THE CHROMATOGRAM

Published procedures for chromatographic separations are, from the
experience of many workers, often difficult to reproduce (2). The
major reason for this is the lack of experimental data recording
in the publication. The fault for this lies in two places, either
with the authors which fail to include such pertinent date in the
manuscript, or with editors which screen such data out before publication to "simplify" the report. All authors should make a
definite effort to report all details.

The following recommendations are presented here as a guide
to the reporting of chromatographic data. The list covers most
situations; a summary of major items of data that should be included in every report follows the detailed list.

Guide To Reporting Chromatographic Data

Tank or chamber

1. Name of manufacturer; internal dimensions.
2. How used--for example, in special chromatography room or
on open bench.
3. Tank sealed by glass contact, grease, etc.
4. Lined with paper or cloth, etc., for atmosphere saturation.
5. Saturated or not, and if so, how and for how long before
chromatography.
6. Temperature of room or chromatography cabinet.
7. Humidity and relative humidity--this is of extreme importance in many separations.

Paper or thin layer

1. Manufacturer of paper or thin layer; whether home-made,
etc.
2. Type of paper or thin layer, including whether channeled
or serrated; dimensions used; foil, plastic, or glass plates;
thickness of layer; wet or dry.

General rules

(1) Spots are outlined with solid or stippled lines according to their intensity and distinctness:

Intensive,   Weak spot with
clear spot   diffuse outlines

(2) Spots are hatched according to their color with the main colors as follows:

Blue   Yellow   Red

Intermediate colors are obtained by "mixing":

Green   Purple   Orange   Brown

Further hues may also be indicated; for example:

Violet   Greenish   Reddish   Greenish   Brownish
              yellow   yellow   blue   yellow

Chroma may be indicated by means of various densities of hatching:

Low: High:
Greenish   Saturated
white   green

UV—absorbing spots are marked with horizontal lines:

General   Colored dark   Brownish
marking   spots, e.g.:   dark
               Reddish dark

Simple rule for memory:

B/ / /   Y| | | R\ \ \

Blue   Yellow   Red

Figure 8.1.   Key for marking the color of fluorescent spots on chromatograms, according to Nybom (1).   Courtesy Journal of Chromatography.

3. Any additives such as fluorescent indicators, buffers, binders, etc.

4. Ascending or descending technique, etc.; pretreated by user prior to chromatography; distance of origin line from edge; length and time of run; whether rerun using same or different conditions.

5. Method of supporting paper or layer; number of chromatograms in the tank--this can have a major effort on separation, and frequently many workers keep blank chromatograms to fill the tank so that this factor remains constant (blanks can be dried and reused).

6. Method of activation, if any, together with subsequent method of storage. This is important also for quantitative work.

Mobile phase

1. Details of each component of the solvent mixture, e.g., petroleum ether (b.p. 100-200°), redistilled, dried or not. If aqueous solutions are used, then the molarity or concentration should be stated.

2. Composition of solvent; monophasic or upper or lower layer of two-phase mixture; whether the lower layer was used for saturation or rejected; give total volumes of solvents used as well as proportions.

3. Equilibration using one or both phases, wetting of internal lining of tank, time for equilibration.

4. Method of drying chromatogram after solvent run.

5. Number of times mobile phase can be used before replacement is necessary.

Sample application

1. Type of applicator; manufacturer's identification.

2. Spot, band, or streak, etc. applied.

3. Solvent used to dissolve solute and method of drying spots or streaks after application.

4. Was the chromatogram handled, or was a "no-touch" method used? Was the whole paper or layer open to the atmosphere, or totally covered with only the origin exposed?

5. Amount of solute and volume of solution applied, and size of spot or streak; analytical or preparative run.

Location reagents and methods

1. Spray, dip, radiation monitoring, etc.; if viewed in UV, then wavelength used; if a photodensitometer was used, then quote wavelengths used, filters, speed, type of record, etc. (live de-

tails of color and stability of colored product, fluorogen, etc.);
time-lapse between visualization and analysis.
    2. Composition of chemical or biological reagent used; sta-
bility of reagent and storage conditions; order of mixing of components of the reagent if important, together with any time factors involved; room temperature or oven heating.
    3. Details of color and other changes that occur with time.
    4. Quote R_f values or, better, relative R_f values, i.e.,
R_x. For sugars, glucose is often used for amino acids glycine or
proline, etc. If identity is claimed with a known compound, was
the known run in parallel or by applying both to the same spot
(isographic), and in how many different mobile phases?

    Additional details to be reported would include details of
derivatives formed, if any. If the mobile phase reacts with the
visualization reagent, the drying procedure should be carefully
and completely described.
    The above condenses to the following items about every chro-
matogram that should be permanently recorded, usually in the research notebook and with the chromatogram itself, accompanied by
a cross reference:

    Date
    Chromatogram Number
    Substances applied to plate; e.g., nature of extract, name
and concentration of known substances.
    Amount of substances applied
    Sorbent layer, including brand name, indicator, special pre-
paration, thickness
    Plate size
    Activation procedure, if any
    Mobile phase systems used
    Length and type of development
    Temperature
    Developing chamber used; type, size
    Drying procedure
    Detection and visualization
    Identifications made and evaluation.

    It is often convenient to prepare a standard or printed form
with space for this information so that it is easily filled in and
filed or stored. A card index of filed or punched cards may also
be useful in that the chromatograms can be categorized in many
ways.
    The actual chromatogram can be preserved or documented in a
number of ways.
    One of the simplest ways to preserve a chromatogram or a por-

tion of one is to press cellophane tape or clear contact paper on
top of the layer such that some excess hangs out from either side
of the plate. A smooth surface, such as a beaker bottom, is rub-
bed on top of the tape so that the adhesive comes into uniform
contact with the layer. As the tape is carefully peeled away, the
upper portion of the layer should be attached, and the whole may
be fastened in a notebook with the excess tape at the edges.

Berlet (3) recorded chromatograms on transparent paper.
After the developed plate has been observed and marked under UV
light or sprayed with a visualization reagent, it is covered with
a clean uncoated plate of equal size. This plate will protect the
layer. A sheet of transparent tracing paper is clamped in place
on top of this with four small spring clips. The spots are traced
onto the paper with any desired notations. A light box may be
located underneath to aid visibility and the location of weakly
reacted spots. This procedure has the advantage that it leaves
the layer intact for any additional analytical procedures such as
in situ quantitation or additional spraying.

8.3 DOCUMENTATION BY PHOTOGRAPHY

Photography is certainly the preferred method for the permanent
recording of chromatograms. Both standard color (4-6) and
Polaroid (7-9) have been used for photography in visible light,
and color (4,7,10,11) and black and white (8,9) have been used for
photography in UV light.

A complete synopsis of photographic documentation of TLC
plates, including details on filters, exposure monitoring, and
experimental photography, with actual photographs of TLC plates
for illustration, prepared by Heinz and Vitek has been written
(12). The worker interested in such documentation should consult
this article.

Caldwell et al. (5) have devised two data sheets for use in
photographing a chromatogram; the first of these is shown in Figure 8.2. Data are entered on the sheet prior to development, with
any additional data or remarks entered during visualization and
evaluation. The sheet is placed adjacent to the chromatogram and
photographed. The sheet is filed in a folder, loose-leaf book,
or notebook for subsequent filing of the photograph.

Figure 8.3 illustrates a small (2.75×8 in.) legend sheet for
use with 20×20 cm plates that have been developed for a distance
of 10 cm or less. Data are placed on the sheet prior to chromatography, and after development and visualization additional comments may be entered before it is placed on the plate to be photographed with it. Caldwell et al. (5) reported the following
photography conditions: visible light; Kodacolor X film, f5.6 for

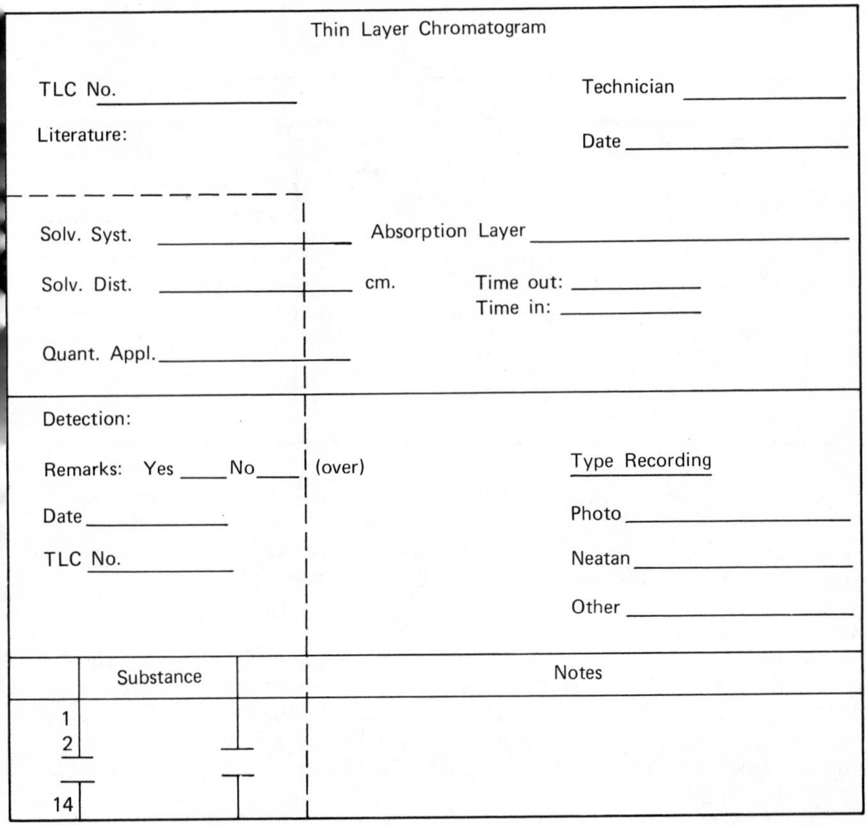

Figure 8.2.   Legend sheet (8.5×11 in.) for use when photographing a 20×20 cm chromatogram.   Portion to left of dashed line is included in photograph of plate (5).   Courtesy Journal of Chromatography.

1/50 sec, two 15-W fluorescent bulbs placed 18 in. above the plate. Ultraviolet light: High Speed Ektachrome film, f8 for 12 sec, two 15-W UV bulbs (General Electric F15T8 BLB, 3200-4000 H) placed 18 in. above the plate.

Xerox copying can also be used to record chromatograms. Felici et al. (13) published a note explaining the conditions under which they obtained satisfactory results. The Xerox model 422 was found to give the best results compared to the other five copiers tested. It was also preferred over Xerox models 3600 and 7000 because of their curved windows. Felici et al. note that

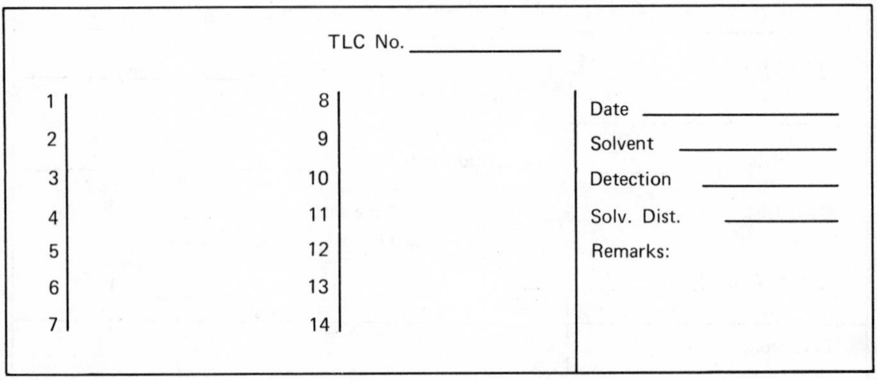

```
                            TLC No. _____

   1 |                  8 |              Date _____
   2 |                  9 |              Solvent _____
   3 |                 10 |              Detection _____
   4 |                 11 |              Solv. Dist.  _____
   5 |                 12 |              Remarks:
   6 |                 13 |
   7 |                 14 |
```

Figure 8.3. Small legend sheet (2.75×8 in) for use when photographing a 20×20 cm chromatogram. When the solvent front is 10
cm or less, this sheet will fit above it on the plate for recording date (5). Courtesy Journal of Chromatography.

Xerox copiers involve reflection of blue/green light and they will
not satisfactorily reproduce blue and green colors. The intensity
setting on the copier apparently had no influence on the copies,
but it was found that intensity could be improved by copying the
copy. However, the background is also intensified, so there is a
limit to this.
      A number of the commercial chromatography suppliers offer
aids for chromatogram documentation. Brinkmann Instruments offer
a photocopier for TLC chromatograms as shown in Figure 8.4. It
can be used to copy visible chromatograms and chromatograms that
absorb UV light at 366 nm; it is not suitable for chromatograms
containing compounds that absorb UV light at 254 nm.
      Brinkmann also offers a binding liquid that dries to form a
flexible plastic coating. This liquid, known as Neatan, is sprayed on the chromatogram and allowed to dry. After it hardens,
transparent tape is placed on top of the layer to peel it off. It
is not suggested for use with home-made coated plates.
      Applied Science Laboratories offers a similar fluid called
Strip-Mix. The film is porous, allowing the extraction from the
sorbent of any substances that are soluble in chloroform or water.
The film may be placed, along with any attached radioactive substances, into scintillation fluid and counted without causing
significant quenching. The following steps are suggested for its
use:

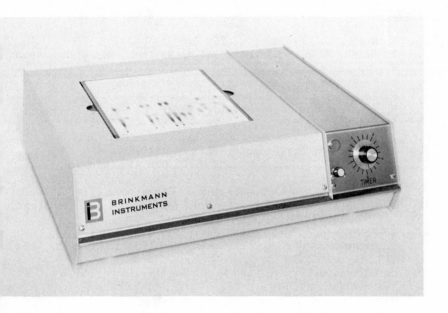

Figure 8.4.   Photocopier for TLC chromatograms.   Courtesy of
Brinkmann Instruments.

    1.   Set the plate on a paper towel with the sorbent layer up.
    2.   Place spacers on opposite sides of the plate.   These may
be any narrow cardboard or metal strips approximately 0.5 mm thick
and as long as the plate side, and are set upon the layer.
    3.   Pour 20-25 ml of Strip-Mix onto one end of the plate, and
spread it evenly across the surface with a glass rod resting on
the spacers.
    4.   Push the excess off one end onto the paper towel.
    5.   Allow the layer to dry 10-15 min.
    6.   Using a razor blade, cut out individual areas as desired
or the entire layer.   These areas should easily peel away.

    The layer thus obtained may be used as a permanent record,
or substances of interest may be extracted for further analysis,
chromatography, radio-counting, etc.

REFERENCES

1. N. Nybom, J. Chromatogr., *26*, 520 (1967).
2. I. Smith, A. D. Baitsholts, A. A. Boulton, and K. Randerath, J. Chromatogr., *82*, 159 (1973).
3. H. H. Berlet, J. Chromatogr., *21*, 485 (1966).
4. R. Jackson, J. Chromatogr., *20*, 410 (1965).
5. R. L. Caldwell, J. K. Shields, and L. S. Stith, J. Chromatogr., *36*, 372 (1968).
6. J. F. Gonnet, J. Chromatogr., *86*, 192 (1973).
7. A. S. Milton, J. Chromatogr., *8*, 417 (1962).
8. E. Hansbury, J. Langham, and D. G. Ott, J. Chromatogr., *9*, 393 (1962).
9. K. Randerath and E. Randerath, J. Chromatogr., *16*, 111 (1964).
10. I. D. Jones, L. S. Bennett, and R. C. White, J. Chromatogr., *30*, 622 (1967).
11. G. P. McSweeney, J. Chromatogr., *33*, 548 (1968).
12. D. E. Heinz and R. K. Vitek, J. Chromatogr.,Sci., *13*, 570 (1975).
13. R. Felici, E. Franco, and M. Cristalli, J. Chromatogr., *90* 208 (1974).

CHAPTER 9

Quantitation

9.1 INTRODUCTION

Concomitant with the availability of spectrodensitometers in recent years, quantitation by evaluation *in situ* is becoming common.
However, there are a number of ways in which quantitation of the
amount of a substance separated on a thin layer chromatogram can
be performed. Densitometry *in situ*, and other methods will be
discussed in detail in this chapter.

Quantitative methods will not be successful unless the sample
and the reference standard are properly handled. The samples must
be kept in closed vials, and the solvents used should preferably
be of low volatility, since evaporation during application procedures can change the concentration. This is particularly true
in routine work when a standard solution must be opened repeatedly
for application of repetitive amounts on many chromatograms. A
Reacti-Vial (Pierce Chemical Co.) with a Teflon stopper through
which the needle of a syringe can be forced for sample withdrawal
seems ideal for overcoming this problem. This is not suitable for
the disposable glass sampling capillaries. Standard solutions
should not be kept too long, as they become concentrated and may
decompose or develop extraneous contaminants.

9.2 VISUAL TECHNIQUES

From estimation by eye many years ago to *in situ* scanning in a
spectrodensitometer, quantitation of substances separated by TLC

has come a long way. Visual evaluation of thin layers can be carried out for quantitative purposes in a number of ways.

In the earliest attempts to quantitate the amount of material
separated on a plate the spot size was visually compared with the
size of a known amount of the equivalent substance separated on
the same chromatogram. Spot areas appear to be a logical first
attempt at quantitation using visual means. The relationship between the spot size and the amount of substance therein is influenced by the sorbent, the mobile phase, and the conditions of the
chamber. For this technique to succeed, good separation is needed, as well as the largest possible change of spot size per unit
of change in the amount of substance separated. At lower concentrations per spot area it is possible to obtain a linear correlation between weight and area. The area of the spot compared
to the density within the area may be the limiting factor here.
It is easier to evaluate spot size than small changes in the density when visual evaluations are being made. Planimetry has been
used to measure the spot size, but this is difficult to do on
friable thin layers of sorbents.

The visual method has seen little use in recent reported ex-
perimentation. However, in some applications visual evaluation
could be all that is necessary. For example, in many quality control procedures a visual scan can show the presence or absence of
impurities in a given preparation.

There are two reasons for using the method:

1. To determine that impurities are within certain limits.
2. To determine the actual number and quantity of the impurities that may be present.

One example of the first use is to apply serial amounts of
the sample in question to the plate and after chromatography evaluate the results visually. Ideally there would be only one spot
present if the solution contained a pure solute. This is the
simplest approach but has many shortcomings. At a low solute concentration only one spot may be seen, whereas if a larger amount
is applied other minor constituents may show up in the sample.
The results may vary from laboratory to laboratory and between two
individual operators in the same laboratory, particularly if
quantitation is desired. Various conditions of plate development
and spot travel may affect the ability to see some solutes,
particularly if they are present in small amounts. The decision
to apply a larger sample or to slightly vary the development conditions may depend upon different situations. Generally, this
approach should not be used if reliable quantitative results are
desired.

The other consideration in this type of study is the reactivity of the detection reagent with the solute of interest and the impurities that may be present. It is not valid to assume that all substances within a class will react to the same degree with the reagent. This is also true when considering *in situ* scanning methods. Calibration curves for quantitation must be prepared for each compound under consideration.

Probably the best way, and the method most often used, is to apply standard quantities of the solutes in question and compare the corresponding spot sizes after development. The use of visual techniques for quantitative purposes requires a knowledge of the nature of the impurities and a source of the pure substance for reference. Also required are well-separated spots. The R_f should be high enough that the spot is not too concentrated. Spots of R_f between 0.3 and 0.7 are more readily evaluated, since the area per unit of solute is more uniform than those of higher or lower R_f's. Spots of low R_f are too concentrated, and visual evaluation of area or intensity are erroneous. Likewise, the areas per unit of mass for spots of high R_f are too diffuse and are therefore difficult to estimate.

There are a number of ways that the spots can be visualized other than by the inherent color of the solute discussed above. The zones can be seen as dark spots against a fluorescent background on the F-254 sorbents. These sorbents contain a phosphor that shows an intense fluorescent background when activated with a short-wave ultraviolet lamp. If it is absorbing near the wavelength of activation of the phosphor, the solute will show up as a dark spot against the brilliant background. The assessment of spot quantity by unaided visual comparison is poor. It is difficult to grade the spot area. The "measles effect" can be a disadvantage. This refers to the phenomenon of seeing extra spots after looking away from the brilliant plate several times or after blinking.

Fluorescent compounds can be detected on thin layers at the nanogram and subnanogram level. Here again, because of visualization in dark rooms under ultraviolet light, problems arise. Pons et al. (1,2) have studied the precision of quantitation of fluorescent aflatoxins separated by TLC. In quantitative measurements of fluorescence intensity, a possible error of 30-50% can occur when an unknown is visually compared to one of two standards of different concentration. An error of 15-25% can occur when an interpolation is made between two such standards, as reported by Pons in earlier work (3).

Johnson (4) has described the results of visual quantitation of $^{4-}$chloroacetanilide in phenacetin. The quantity of sample that amounted to 0.2 μg of phenacetin contained 0.01% 4-chloroacetin as impurity. The results of the collaborative study of the use of visual examination of the layers after separation is shown in

Table 9.1. As seen in the table, there was only fair agreement
among the different laboratories. This type of assessment of
thin layer chromatograms is subject to errors of 20% or more. As
a general evaluation technique, the procedure may be suitable for
routine quality control, but for precise quantitative purposes
visual examination is not always applicable.

TABLE 9.1 PERCENTAGE OF 4-CHLOROACETANILIDE FOUND IN
SAMPLE OF PHENACETIN QUANTITATED VISUALLY (4)

Sample No.	Laboratory				
	A	B	C	D	E
1	<0.01	<0.01	<0.01	0.005	--
2	0.01	--	0.01	--	0.02
3	0.02	0.04	0.02	0.03	--
	0.03	0.03			
4	0.07	0.05	0.07	0.07	0.06
5	0.04	0.08	0.07	0.06	--
	0.05				

9.3 EVALUATION BY MEASURING SPOT AREAS

After separation of the solute or solutes in question, the quantitation can be performed by measuring the spot area. This method
has been applied intensively in paper chromatography. It is probably easier to adapt the technique to paper chromatograms than to
TLC, since the paper is more uniform than thin layers and the
spots may in general be more discrete. However, the method has
been extended to thin layers by many investigators. A number of
variations of the general concept have been used; no one method
seems to be universally used.

Whether the relationship of area to concentration can be
used for quantitation depends on a number of properties of the
chromatogram. Among these are:

1. Reproducible sorbent characteristics which define activity and characteristics of the sorbent that are used in the layer.
Modern commercial layers seem to fulfill this requirement. They
generally are uniform in thickness and give reproducible chromato-

grams and spot characteristics if the conditions can be reproduced.

    2.   Reproducibility of the application of the sample and the
standard. Since the standards are applied at the same time as
the sample, this is usually possible.

    3.   Use of a developing system that will separate the sol-
utes as discrete and well-separated zones with no tailing.

    4.   Completeness of reaction of the solutes with the detec-
tion reagent. Dipping in the reagent is preferred to spraying,
because it is easier to reproduce. However, the dipping procedures depend on the solvent-solute characteristics. Also, if
aqueous sprays are used, care must be taken not to spray too
heavily or the layer will be removed from its support. (See the
section on application of detection reagents in Chapter 7.)

    If the zones are distinctly outlined, they can be circum-
scribed with a sharp instrument so that there is a permanent marking of the area. This should be done at the same time for each
chromatogram in order to obtain better reproducibility, since
time, atmospheric conditions, and other factors can cause fading
or reduction of the spots. After they are marked, the areas can
be measured in a number of ways. The easiest way is to use a
planimeter, as was done by Oswald and Flück (5). Another way to
measure spot size or evaluate the results is to compare to areas
of the standards visually, as described by Morrison and Chatten
(6) and as noted earlier in this chapter.

    Gänshirt and Poderman (7) described a procedure in which the
spot areas were traced on writing paper, following which the spots
were cut out and weighed. An adaptation of this is to trace the
areas on squared paper (graph paper) and count the number of
square millimeters. Since the areas are usually small, considerable error is present in these measurements.

    Still another method of measuring area is to photograph the
spots and then cut out the zones on the photograph and weigh them,
as described by Schierf and Wood (8).

    To avoid preparing a calibration curve for each plate, a
function expressing a linear relation between the amount of the
solute applied and the spot area can be determined. Brenner et al.
(9) have discussed application of this method to data obtained
from thin layer chromatograms. Spot areas should be proportional
to the logarithm of the solute amount under defined conditions.
Petrowitz (10) determined tar oils, DDT, and γ-BHC within narrow
limits using these relationships.

    According to Purdy and Truter (11), the average deviations
obtained in their work and others was 3.1% for 600 separations
based on adsorption methods and 3.6% for 980 partition chromatograms. The systematic error is small, since the comparison is al-

ways made with the values of the standard solute, which should always be chromatographed on the same plate. Random errors are high because of variations in spot shape and the difficulty of defining the spot area. They can be pronounced in comparisons of results between plates or results obtained on different days.

In spite of the drawbacks mentioned, it is possible to use these simple methods to quantitate if the limitations are understood. With the uniform plates commercially available, it is easier to reproduce chromatographic results if ambient conditions can be controlled. Prerequisite for success in these methods is the use of minimal amounts of solute. Generally it has been found that practitioners of thin layer chromatography overload the chromatogram with solute. Linear calibration curves can be more readily obtained when the spot areas are not overloaded. Solute zones that are too concentrated generally prevent complete reaction with the detection reagent. Furthermore, dark zones are more difficult to assess visually than a series of zones with varying degrees of intensity commensurate with the spot size.

9.4 ELUTION TECHNIQUES

The most widely used of all methods of quantitation of substances separated on thin layers is that of elution. The method of choice in these procedures depends almost entirely upon the nature of the solute separated as well as the sorbent used. A knowledge of the interplay between the two is very important for success in applying elution techniques. The basic problem is removal of the solute from the active sites on the sorbent surfaces. Often low recoveries are reported because these factors have not been considered. The selection of the solvent for elution is also important. After elution is accomplished the identification or quantitation of the solute can be performed by a variety of means. Depending on the nature of the compound separated, there are a number of advantages and disadvantages in detection for elution as well as the means of the actual quantitation. The focus in this section will be the chromatographic and elution problems. The compounds of interest will determine the mode of quantitation.

Before the solute can be eluted it must be located. This is simple if the compound is colored or fluoresces in ultraviolet light. However, this location procedure is not possible with the majority of compounds. They must be located by use of a reference compound separated on the same plate, which is visualized. Preferably the reference compound is spotted on the outer zones and in the middle zone of the chromatogram. This enables one to correct for uneven mobile phase flow. If reference compounds are not available, the solute in question can be applied to the chromato-

ram and then located by use of the appropriate detecting re-
gent, acting as its own reference for locating the zones to be
luted.

It is good practice to mark the solvent front at the end of
development. Usually the line (zones) of separated solute will
e parallel to the solvent line. This can aid in the marking of
he zones to be scraped from the support. After the chromatogram
as been developed, several procedures can be used to locate the
eference zones (spot where solute separated).

Some compounds, particularly if they absorb strongly in the
ltraviolet regions, can readily be detected by exposing the chro-
atogram in the dark to an ultraviolet lamp. Often the amount of
he compound must be rather high before the zones can be visual-
zed. When marking the region to be eluted it is recommended that
n area larger than the visualized zone itself be scribed into the
orbent to outline the area to be scraped off. This method also
as an advantage as a preliminary screening, since the absorption
n the ultraviolet region can be characteristic of the compound
f interest. However, advantage can be taken of the absorptive
haracteristics of the compound when the chromatogram is developed
n a sorbent containing phosphor. Under ultraviolet light the
ompounds that absorb in the ultraviolet region will appear as
ark blue zones against the fluorescent background of the layer.
he zones then can be marked into areas to be removed from the
acking by scraping. The use of this technique is limited to
hose instances where traces of the phosphor will not interfere
ith the analytical technique used for the quantitation. This
ethod is a nondestructive technique and is often used, since com-
ercially prepared plates are available with the phosphor incor-
orated in the sorbent layer.

The methods described below are applicable to the majority of
hin layer chromatograms. Probably the easiest method for locat-
ng a zone, particularly in the case of lipids or lipophilic sub-
tances, is to spray the chromatogram with water. The detection
epends on the differential wetting properties of the sorbent and
olute. The detection is easiest in oblique or transmitted light.
t is generally an insensitive method, but for the reference com-
ound this is not a matter of concern. It has been widely used in
ipid work and occasionally to detect steroids (11,12). Varon et
l. (13) have found the method to be sensitive for the location of
strogen acetates. This method is described in more detail in
hapter 7 under nondestructive detection methods. The technique
an also be used on the solute itself, since it is nondestructive,
nd a reference compound may not be necessary except for use in
ocation of the other compounds of interest.

A common nondestructive procedure is the exposure of the
hromatogram to iodine vapors (14,15). The mechanism of this test

is not known. It may be due to differential solubility of the
iodine in the solute, formation of adducts, or the addition of
iodine to double bonds present in the molecule. These reactions
depend on the type of compound present. It is good practice to
test this procedure on a chromatogram before using it in quantitative work to be sure there is no permanent reaction of the reagent with the compounds.

Iodine crystals are placed in the bottom of a tank, and after
some equilibration the chromatogram is placed in the tank. Usually the reaction is immediate. Yellow to brown spots are observed. After the spots are visualized, the zones are immediately marked by scribing the area with a sharp instrument. If reaction with the areas to be scraped must be avoided, the section
of the plate containing the reference can be cut off and exposed
to the iodine. As the iodine evaporates from the layer, the color
eventually disappears.

Quantitation of zones may also be accomplished by spraying
the TLC with a color reagent, eluting the colored product, and determining it spectrophotometrically. This method, although successful in some cases, is beset by difficulties. Spraying is difficult to reproduce, as the amount of reagent varies in different
areas of the plate. The ratio of reagent to amount of solute in
the spot will also vary among different areas of the spot. Sometimes it may be more difficult to elute the colored compound than
the original substance. Conversely, it is sometimes easier, as
described in the next paragraph.

An example of elution of a zone after spraying with a reagent
for color development was reported by Haefelfinger (16). Quantitation of diallyl-bis-nortoxiferine-dichloride was carried out
by color development with bromothymol blue. Apparently in this
case it was easier to elute the alkaloid from the plate after
chromatography and after complexing with the dye. A twofold
result was obtained: easier elution of the solute as well as a
means of quantitating it. After elution, spectrophotometry or,
as in this case, optical rotation, can be used to assess the
quantitative aspects.

As pointed out by Court (17), the boundary effect must be
avoided. Figure 9.1 illustrates this phenomenon. In spite of
the fact that the solvent front had moved in a uniform horizontal
line up the plate, the row of spots or line of the separated solute may not be parallel to the solvent front. Sometimes it is
possible, particularly with biological samples, to see other solutes, either fluorescent or absorbing under an ultraviolet lamp.
These solutes will form a line that can serve as a guide to mark
the zone to be scraped off. One can also overcome the boundary
effect by including a reference in the middle region of the chromatogram, as described before.

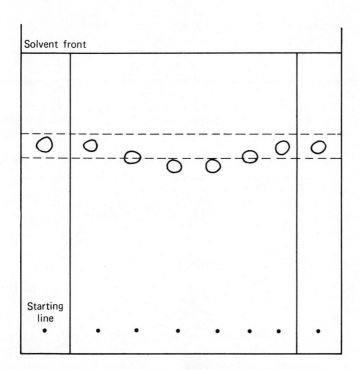

Figure 9.1.   Illustration of the boundary effect.   Outer lanes
contain the reference compounds to denote the zone to be removed.
Note that if parallel lines scribed by position of the reference
are followed, the zones in the middle of the chromatogram will be
missed.

9.4.1   Removal of the Sorbent From the Plate

After marking the zone containing the compound to be assayed, the
next step is to remove the sorbent and transfer it to a suitable
vessel for elution with the proper solvent.   The areas can be
scraped off with a razor blade or straight-edged spatula.   The
sorbent can be collected on a piece of weighing paper or a similar
nonabsorbent sheet, such as cellophane.   From this the sorbent is
shaken into the tube or vessel for elution.
      There are many problems related to this part of the elution
process.   Air currents must be absent, or small particles will be
blown away and result in loss of sample.   Some layers have a hard
surface, and care must be taken to prevent loss of the solute due
to flaking.   When many samples are to be processed, this technique

can be tedious and subject to mechanical loss.

There are a number of suction devices on the market designed to avoid the problems of scraping the sorbent from the layers. Typical of these is the all-glass suction apparatus pictured in Figure 9.2. The inlet tube of the apparatus is positioned close to the layer within the marked zone of interest. The vacuum takes in the sorbent and deposits it on the sintered glass disk. Two o three small drops of water are added to inactivate the sorbent, particularly if it is silica gel or alumina. Pure distilled solvent for elution is then drawn through the nozzle to elute the solute out of the sorbent and into the attached collection flask. Further operations, such as concentration, evaporation, rechromatography, and so forth may then be performed. This type of apparatus avoids mechanical losses found in scraping. The vacuum extraction of Matthews et al. (18) is an example of this.

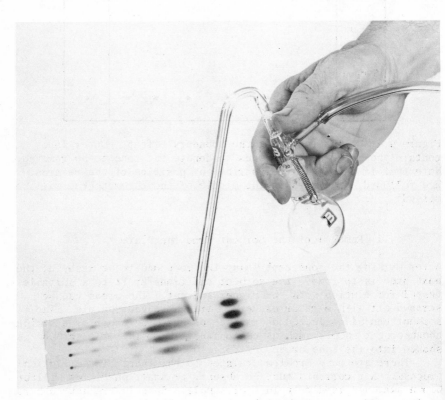

Figure 9.2. Typical suction device used to remove sorbents from the support for elution.

9.4.2 Elution of the Solute From the Sorbent

After addition of the appropriate solvent, the solute may be eluted by simple agitation of the test tube or container, followed by
centrifugation to remove the sorbent. The solvent is then de-
canted, and the procedure is repeated as many times as necessary
to obtain complete recovery of the solute. The elution can sometimes be accomplished with a single extraction of the sorbent. A
small column containing the sorbent can also be used. The solvent
can be added to the top of the column and the effluent collected
in the appropriate vessel for the quantitative step. Lehmann et
al. (19) described a microcolumn to be used under pressure. Attal
et al. (20) used a 3-ml syringe with a 25-gauge needle. A plug
of glass wool in the bottom of the syringe served as support for
the sorbent. The syringe acting as a column is very practical.
Elution methods are given in detail in Chapters 12 and 14.
     The selection of the sluting solvent itself is important,
especially when quantitation is desired. Completeness of elution
will be determined by the solvent. The selection of the solvent
is governed by the nature of the solute and of the sorbent. Constant percentage or completeness of recovery is desirable. Recovery can be assessed by using radioactively labeled reference
compounds. This is particularly important when working with biological samples when the expected amount is not known. Otherwise,
recovery can be determined by the amount of a known substance that
was eluted from the sorbent.
     Aqueous solvents are not suitable for some elutions, since
calcium sulfate present in many TLC sorbents is slightly soluble
in water. Sometimes polar solvents such as ethanol extract other
constituents of the layers. If the R_f of the solute is over 0.8,
the developing solvent or its constituents can generally be used
for elution.
     Deactivation of the sorbent by addition of enough water to
wet the small amount present has been found to increase recovery
of solutes using nonpolar solvents.
     Selection of the solvent also must be considered from the
standpoint of the quantitative method to be subsequently used. If
spectrometry in the ultraviolet is to be used, the solvent must
absorb only weakly; otherwise the solvent must be evaporated and
the solute dissolved in a more appropriate solvent.
     Fine particles of sorbent that can scatter light must be re-
moved by centrifugation or filtration. Filtering through micro-
pore or membrane filters has been found useful for this purpose.
Clear infrared spectra and mass spectra have been obtained after
filtration of solutes removed from silica gel through 4-μ Milli-
pore membrane filters (21).

That deactivation of the sorbent is an important factor in completeness of elution by solvents is illustrated by the experiments from the authors' laboratory summarized in Table 9.2. Radioactive deoxycorticosterone in equal amounts was applied to four zones of a silica gel GF-254 plate. The chromatogram was not developed. The zones were located under an ultraviolet lamp and marked. One zone was scraped directly into a scintillation vial; the second was wetted with water and scraped into another vial. One of the other two was scraped dry into a test tube and eluted twice with methanol. The fourth zone was wetted with water, scraped into a test tube, and also extracted twice with methanol and the extract transferred to the counting vial. Scintillation fluid was added to all four and counting was done. In the first two, the gel remained at the bottom of the vial and the scintillation fluid extracted the solute. Recoveries stated in the table were different. Ganjam et al. (22) also point out some of the problems related to elution of steroid from TLC plates. They went through considerable investigation to find a solvent that would elute cortisol and corticosterone from silica gel. Table 9.3 summarizes the results and also points out the well-known inefficiency in using the same solvent for eluting a variety of compounds with varying polarities and chemical structures even in the same class of compounds.

TABLE 9.2   RECOVERY OF DEOXYCORTICOSTERONE-[14] FROM
SILICA GEL LAYERS

1.	Scraped dry into vial, scintillator then added	37%
2.	Scraped wet into vial, scintillator then added	92%
3.	Eluted (dry) 2 × with methanol	63%
4.	Eluted (wet) 2 × with methanol	64%

Impurities extracted from the sorbent probably are one of the largest sources of error when spectrophotometry is to be used for quantitation of eluted solutes. Silica gel and alumina contain inorganic impurities such as iron and chloride. Depending on how long the plates have been stored, many organic impurities can be absorbed from air pollutants, especially in an active organic laboratory. Impurities may also be absorbed from the packing material of the containers in which the commercial plates are supplied. Some of these problems can be avoided by using preliminary purification procedures. If the plates are prepared in the laboratory, the gel can be washed with dilute sulfuric acid or

TABLE 9.3   COMPARISON OF DIFFERENT TECHNIQUES FOR ELUTING
CORTISOL AND CORTICOSTERONE FROM SILICA GEL

Elution method	Percent recovered*	
	Cortisol	Corticosterone
Partition of silica gel between benzene and water	1 ± 0.2	96 ± 1.3
Dichloromethane:methanol (9:1)	90 ± 1.3	92 ± 1.4
Cold methanol	16 ± 1.0	62 ± 1.0
Hot methanol	25 ± 1.0	65 ± 0.8
Cold ethanol	15 ± 1.0	55 ± 1.8
Hot ethanol	15 ± 1.3	60 ± 1.0
Chloroform:methanol (1:1)	33 ± 1.4	50 ± 1.0
Column chromatography	25 ± 1.3	71 ± 1.2

* Each value represents the mean ± the standard error of eight
  determinations.

hydrochloric acid (23). Ethanolic hydrochloric acid has been
used for washing the silica gel (24). Problems arise here, since
these washing procedures may remove unknown amounts of binding
material. Fine sorbent particles that are necessary for adhesion
of the layer are removed, and furthermore new impurities may be
added.

Sometimes commercial layers can be rid of extraneous material by development with the solvent to be used for elution. Methanol has been used in this way. If the solvent is allowed to
overrun the plate in the developing tank, impurities are washed to
the top of the plate. It is not unusual to see a yellow line at
the solvent front and eventually a band of yellow at the top of
the plate.

A unique way to avoid removal of the sorbent from the chromatogram and still allow elution *in situ* has recently been described by Falk and Krummen (25). The method requires a special
instrument designed for the purpose and is entirely automatic.
The use of this instrument, available commercially, is described
below.

One to six samples are applied, evenly spaced, on TLC plates,
and developed in the usual manner. A milling device is used to
remove the sorbent layer around the spot to be recovered. This

permits a tight seal between the backing of the plate and the
Teflon elution head. The elution head is then placed over the
milled-out area and pressed firmly into place with a clamping bar.

The syringe reservoirs are filled with the same solvent or
with different solvents. The precise amount of solvent to be used
for elution is preset, and the rest is automatic. From 1 to 5 ml
of solvent flows from the syringe reservoirs through the elution
head at a preselected rate of speed and into the cuvet. As the
solvent flows through the elution head it flows through the sorbent and lifts the solute from the layer. When the preset volume
is reached, the instrument automatically turns off. No inter-
vention is required during the elution process, unless a different
volume is desired for elution for each zone.

When the elution is complete, a pump forces air through the
elution head to drive out the remaining solvent and to dry the
layer under the elution head. The clamping bar is released and
the elution heads are removed. The resulting solution collected
in the cuvet or volumetric flask is free of sorbent and ready for
analysis.

The minimum amount of solvent that can be used is approximately 0.75 ml. This, of course, depends on the amount of solute
present and the elution power of the solvent and/or solubilities.
Also, the speed of elution is important. The slower the flow of
solvent, the better the elution. If set at a fast speed, larger
amounts of solvent are required because it then becomes more of a
washing effect, rather than the desired adsorption/desorption process. At the slowest possible elution speed it takes 10 min for
each 1 ml of solvent.

The utility of this apparatus and its applicability are dependent on the quantitative removal of the solute from the sorbent. It is not always possible to obtain quantitative removal
of solutes. Therefore, this type of instrumentation will have
some limitations.

It presents one solution to the ever-present elution problem.
The instrument is limited to six zones. The eluting heads are
very wide, and it is necessary to have the desired solute well
separated from other materials or else scrape away the unwanted
material. It is also not adaptable to eluting solutes that have
been streaked the length of the origin of the chromatogram. This
instrument is described in more detail in Chapter 14 in connection
with other preparative techniques. Table 9.4 shows the recoveries
obtained on eluting different compounds using this procedure.

9.5 METHODS FOR MEASUREMENT AFTER ELUTION

The ultimate goal in quantitative analysis is to determine the

TABLE 9.4 RECOVERIES OBTAINED BY AUTOMATIC ELUTION

Substance	Sorbent	Amount Applied (mg)	Solvent	Percentage Recovery	
				1st Fraction 0.5 ml	2nd Fraction 0.25 ml
Caffeine	Silica gel 60 F-254	36	Chloroform 90 Ethanol 10	100.5	0.25
Benzocaine	Silica gel F-254, basic	13.5	Chloroform 95 Ethanol 5	100.6	0.25
Phenacetone	Silica gel GF-254	18.4	Chloroform 95 Methanol 5 Conc. ammonia 0.15	99.2	0.25

249

amount of a particular substance in a sample. When the procedure
is based on quantitation by some characteristic other than weight,
known amounts of pure reference material must be processed in the
same way to obtain calibration curves for interpolation of the
amount in the unknown. These observations also give information
on the statistical accuracy of the method.

Quantitation by gravimetric determination, in spite of its
poor accuracy, is the only technique that does not require calibration when enough solute or unidentified substance is available.
This method can be advantageous. It has been used for determination of the components of cabbage wax by Purdy and Truter (11).
Here again, once is faced with the problem of quantitative removal
of the solute from the sorbent.

Gravimetric analysis is usually included as part of the process of preparative thin layer chromatography. Sometimes a number of chromatograms are processed together, and the desired solute is collected from the layers and eluted in a single operation.
To find the area where the desired component has migrated, reference substances must be applied to the chromatograms and used
for guides to locate the zones. This procedure is described in
detail in preceding sections of this chapter. After elution the
solvents are evaporated, and after careful drying the sample is
weighed. Since the majority of work in TLC involves micro samples, the use of gravimetric procedures has not been widespread.

9.5.1 Spectrophotometric Methods

Most of the quantitative evaluation of samples eluted from TLC
plates has been done by spectrophotometry. The extract is made up
to a standard volume, and spectrophotometry at the desired wavelength is performed. Many compounds will absorb in the ultraviolet region, and advantage can be taken of these absorption properties; others are colored, and a determination can be made using
the maximal wavelength of the chromophore. If the solute has none
of these properties and is colorless, it must be submitted to a
chromogenic reaction. There are a great number of reagents that
may be used to form a colored derivative after reaction with the
sample in question. Many suitable reagents for this purpose are
discussed in Chapter 7. Often some of these reactions will result
in the determination becoming more sensitive, as in the case of
the reaction of dansyl chloride (26) or Fluram (27), which form
fluorescent derivatives. With amino acids and other amines, the
formation of fluorescent derivatives provides a method of increasing the sensitivity as much as a hundredfold, as well as increasing the specificity of the determination.

Some difficulties are encountered in spectrophotometry of
material eluted from thin layer chromatograms. These are inherent

in the sorbent. Some substances that absorb very strongly in the
regions below 250 nm are eluted from the sorbent, particularly
from silica gel and alumina. These impurities show a nonspecific
absorption spectrum. The intensity falls steadily as the wavelength increases and becomes negligible as the wavelength reaches
400 nm. Some of the impurities may be those washed to the upper
levels of the chromatogram during development and can be seen by
viewing the chromatogram in a dark room under a short-wave ultraviolet lamp.

One of the ways to minimize these impurities is to use as
the blank or control the extract of a blank portion of the sorbent
equal in area to that of the sample and from a zone of equivalent
R_f. This must be processed exactly as the sample and then used
as a blank in the reference beam of the spectrophotometer. The
contribution of the blank can be a major proportion of the absorbance and varies widely with the source of the sorbent and the
mobile phase used for development of the chromatogram as well as
the solvent used for the extraction of the solute.

There are other analytical methods available for quantitation
and identification of the solutes eluted from plates. Among these
are gas chromatography, infrared spectrometry, and nuclear magnetic resonance. These techniques are outside the scope of this
book, although some of them are in Chapter 14.

9.6 *IN SITU* QUANTITATION BY SCANNING

Quantitative TLC analysis involving elution before measurement is
time-consuming and laborious. Although measurement *in situ* as a
simpler method has been generally accepted, the theoretical basis
of such measurements is apparently little known. On the other
hand, recent progress in measuring techniques make direct quantitative TLC evaluation attractive and applicable to many areas of
research and routine work, if the same precision is maintained as
with elution methods that involve spectrophotometry of solutions
of the eluted solutes. Many *in situ* methods using densitometers
and spectrodensitometers are now available, and these can be very
precise, having a relative standard deviation of less than 5%.
Frequent reports indicate that it is possible to obtain precision
with standard deviations below 1%.

Since the early work of Privett and Black (28), Jork (29),
and Barrett et al. (30), using the densitometers then existing,
many instruments have been designed for *in situ* evaluation of thin
layer chromatograms. Some of these instruments represent accessories to existing spectrophotometers. Others are spectrodensitometers designed for just this type of *in situ* scanning. These
instruments also have the capability of scanning gels from elec-

trophoretic separations.

Basically, the optical systems of all of these instruments can be categorized into the designs shown in Figure 9.3. They all have a light source, condensing and focusing systems, and a photosensing detector. Some of them have monochromators, whereas others rely on filters for selection of proper wavelengths for operation. Few of them have monochromators for both the exciting as well as the emitted light in fluorescence.

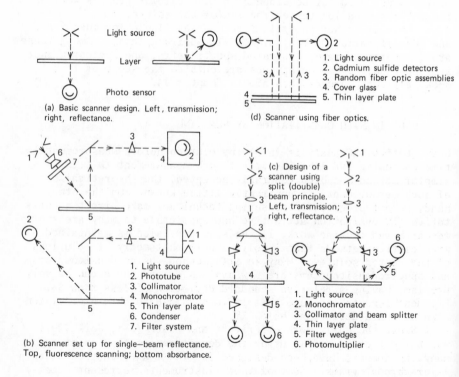

(a) Basic scanner design. Left, transmission; right, reflectance.

(d) Scanner using fiber optics.

1. Light source
2. Cadmium sulfide detectors
3. Random fiber optic assemblies
4. Cover glass
5. Thin layer plate

(b) Scanner set up for single—beam reflectance. Top, fluorescence scanning; bottom absorbance.

1. Light source
2. Phototube
3. Collimator
4. Monochromator
5. Thin layer plate
6. Condenser
7. Filter system

(c) Design of a scanner using split (double) beam principle. Left, transmission; right, reflectance.

1. Light source
2. Monochromator
3. Collimator and beam splitter
4. Thin layer plate
5. Filter wedges
6. Photomultiplier

Figure 9.3.   How a densitometer works.

Probably the simplest densitometer can be described as shown in Figure 9.3(a). This design is embodied in some of the earlier instruments that were originally designed to evaluate separated zones in gels from electrophoretic separations. Some of these have been used with limited success in scanning thin layer chromatograms.

In the true sense these are densitometers, since they are rarely equipped with the adaptations to provide monochromatic light and are usually constructed to scan in the transmission mode. The light source can be either above or below the thin layer, with the photosensor on the opposite side for transmission or on the same side for reflectance. The light source can be a hydrogen, mercury, high pressure mercury-xenon, or xenon lamp for ultraviolet light or an incandescent lamp for visible light.

Figure 9.3(b) is a simple block diagram of an arrangement that can be used in accessories to be attached to available spectrophotometers. In the lower diagram the light from source 1 passes through a monochromator 4 where the desired wavelength is selected and condensed in lens 3 to be reflected from the mirror and directed to the thin layer 5. The reflected light is then viewed by the phototube at position 2 stationed at a 45° angle to the plate. Alternatively in this system, the lamp can be positioned at 1, with the light passing through condenser 6 and filter system 7 before striking the layer at a 45° angle. The perpendicularly reflected light is caught by a deviating mirror and directed through a collimator, 3. The light then passes through the monochromator, 4, to isolate the desired wavelength and detection is by the photomultiplier, 2, which has been moved behind the monochromator. This is shown by the upper diagram in Figure 9.3(b). This instrument was designed to follow reflectance determinations.

Neither of the designs described in Figure 9.3(a) and (b) has provisions for scanning in the double-beam mode. It has been pointed out by a number of authors that this mode is necessary for precise quantitation in scanning of thin layers because of background noise. The instrumental diagrams in Figure 9.3(c) and (d) were designed to provide double-beam operation.

Figure 9.3(c) is a diagram of an instrument designed for both double- and single-beam operation. It can be used in the transmission or reflectance modes as denoted in the diagrams. It has a beam-dividing system, which provides one beam for the reference part of the layer and another for the sample scanning.

Figure 9.3(d) describes the basic design of an instrument that presents a departure from the usual method of carrying the light to the layer. Glass fiber optics are used to transmit the light from the source directly to the plate.

Two randomly gathered fiber optic light pipes with 2-cm^2 heads are mounted side by side so that the beams of light from the source will be carried through the fibers perpendicularly to the layer. The reflected light is passed back through the reverse fiber and directed to the photosensors. The scanning head is close to the layer and allows it to collect most of the reflected light, since the loss of scattered light is now at a minimum.

In the discussion so far, only optical layouts have been described. Recorders to visualize the scans and integrating devices are generally included as parts of the list of equipment needed to perform quantitative work. If a monochromator is not used, filters should be available. The provisions for scanning include (1) movement of the plate over or under the light beam with a stationary viewing photomultiplier or (2) movement of the light beam along with the enclosed viewer while the plate is stationary. Most instruments have the moving plate design. Recording charts can be synchronized with the speed of the plate.

The diagram does not show how the photomultiplier response is handled. Some instruments use only an indicating meter calibrated in absorbance units. Others in double-beam operation amplify and then feed the signal into logarithmic converters. At the output the reference signal is subtracted from the sample signal and fed into an integrator. From there the signal goes to a conventional line recorder, which records the ratio of the two beam signals.

There are a number of modes of operation of these instruments, but not all modes are available on all instruments. All the instruments have some of the operating parameters, and selection of an instrument should be based on the use intended for it. The most versatile instruments have the following capabilities:

1. Scanning in the transmission mode.
2. Scanning in the reflectance mode.
3. Fluorescence scanning.

   a. Transmission mode
   b. Reflectance mode

4. Fluorescence quenching evaluation.

The choice of mode depends on the compound in question and the nature of the sorbent.

Before one can determine the capabilities of the instrument on hand, a few basic principles of the theory and instrument variables should be understood. Calibration of the instrument in the different modes is an important prerequisite. Novacek has reviewed the theoretical and practical aspects of densitometry as it is practiced today (31).

Present methods are frequently based upon empirical calibra-

tion. To achieve linearity, absorbance of various functions of
transmission and reflectance are plotted versus weight or function of weight of the light-absorbing compounds. Therefore, the
question often arises as to whether to prefer measurements using
transmission or reflectance techniques.

The following is a review of the underlying theory for transmission and reflection of light in highly scattering media. Theoretical conclusions, for example, that transmission is to be preferred to reflectance, are discussed with regard to measuring
methods used with presently available densitometric instrument
systems.

The effects of instrument variables on measurement accuracy
for the determination of colorless UV-absorbing compounds by fluorescence quenching and reflectance and the quantitation of colored compounds are discussed in terms of response-peak areas versus
amount spotted.

The theory of radiative transfer in scattering media is due
to Chandrasekhar (32). However, the transport equation has no
analytical solution, and therefore simplified expressions are
employed. The most widely used is that of Kubelka and Munk (33,
34). The work of Goldman and Goodall (35), based on the Kubelka-
Munk theory, has shown that on theoretical grounds measurement
using transmission is more advantageous than that using reflectance. These workers have also proved this experimentally. After
examining the conditions under which the Kubelka-Munk equations
hold true, they concluded:

1. The entering light need not be exactly parallel and perpendicular to the specimen; however, the angular distribution of
incident light should be kept reasonably constant.
2. Monochromaticity of the incident light is important.
Uniform absorption by the sample within the applied spectral band
width is necessary to comply with Beer's law.
3. The illuminating aperture must be small so as to cover a
limited range of absorptions. This range depends upon the average absorption of the measured area and the precision required.
4. Nonuniformity of the sample spot through the depth of the
layer is likely to have a greater effect on measurements using
reflectance than upon those in which transmission modes are used.
5. Variations in the scattering coefficient over the area
of illumination, through the depth of the layer and over the wavelength range, are likely to be smaller than the corresponding
variation in absorption.

Using these results and the Kubelka-Munk formulas, Goldman and
Goodall computed the values for absorbance in transmittance and
in reflectance relative to the background for a wide range of

variables including coefficients of absorption in relation to
layer thickness. The results are shown in Figures 9.4 and 9.5.
From the date shown in the figures, it is obvious that:

1. Linearity of response to absorption of light by a zone of
a layer is more closely approached in the transmission mode than
in the reflectance mode.

2. The response in reflectance is less than the Beer's law
slope (0.4343 in this case), whereas in transmission the response
is greater. This means that for the same sample spot the scattering power of the medium substantially increases the recorded
peak area obtained in transmission and decreases the recorded peak
area in reflectance.

9.7 DENSITOMETER USE

There are, as we have seen, four different ways by which a chromatogram can be scanned with existing densitometers for solutes
separated on thin layers. The densitometers are operated either

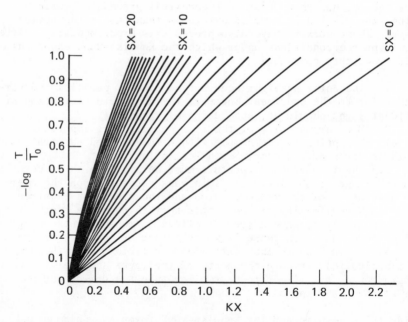

Figure 9.4. Absorbance as measured in the transmission mode.
   KX = absorption relative to layer thickness
   SX = scatter relative to layer thickness

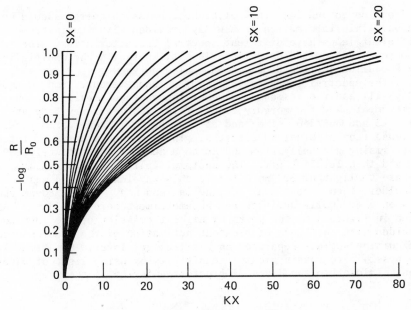

Figure 9.5.   Absorbance as measured by reflectance.
    KX = absorption relative to layer thickness
    SX = scatter relative to layer thickness

in the transmission or reflectance mode. Colored compounds or
colorless non-UV absorbing compounds (derivatized) can be quantitated by the determination of absorption in the visible wavelengths, either by reflectance or transmission. Colorless materials that absorb UV light can be determined by measuring the absorption of the incident UV radiation in the reflectance mode.
When the sorbent contains phosphor, colorless compounds that absorb in the UV regions also can be quantitated using measurements
based upon fluorescence quenching. This technique can be carried
out in either the transmission or reflectance mode. Fluorescence
can also be used for the quantitation of minute amounts of substance that are fluorescent or have been made fluorescent.

9.7.1 Instrument Variables

Some understanding of instrument variables is necessary for precision work in scanning thin layers. Touchstone et al. (36) have
made a critical evaluation of various operating modes of a double-
beam densitometer.
    The properties of the light source in the instrument should

be considered and the lines of highest emission energy should be
known. This information is usually provided with the instrument.
Different lamps have different spectral characteristics. The response curves of the detecting element, whether a photomultiplier
or a cadmium sulfide diode, should also be known.

The spectral characteristics of a xenon-mercury lamp (Hanovia
901.B-11) and the response curve of the photomultipliers in one
instrument show the mercury bands of the xenon-mercury lamp at
312/313 and 365/366. The band at 365/366 nm is considerably more
intense than at other wavelengths. The lamp has a curve of intensity rising gradually from 200 nm to a hump at approximately 500
nm and then leveling off. The photomultipliers have a maximum response to light of 380 nm. Before this wavelength there was a
shoulder, then a continual decline in response. Above 380 nm the
response gradually declined and became almost zero at 570 nm. As
seen in Figure 9.6, the peaks of highest emission of the lamp co-
incided with the bands of greatest activation of the fluorescence
of quinine sulfate separated on a silica gel layer. Sometimes it
is possible to obtain greater sensitivity by using light of wavelength closer to the lines of greatest emission of the lamp.

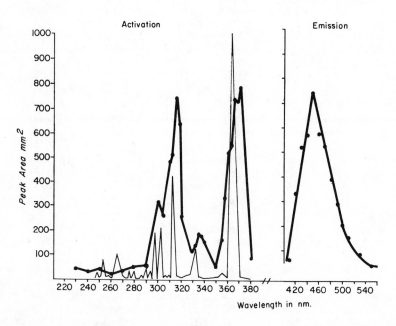

Figure 9.6. Activation and emission spectra of quinine sulfate
(dark line) also shown are the peaks of maximal energy of the
light source (light line).

Figure 9.7 illustrates the importance in quantitative work of operating at the peak of absorption of excitation bands. In the case of 2,7-dichlorofluorescein, scanning away from the maximum of excitation (315 nm) resulted in nonlinear curves. Scanning at 315 nm gave a linear calibration curve.

The importance of scanning at the wavelength of maximum absorption of the solute in the sorbent is illustrated in Table 9.5. The maximum of absorption of a compound adsorbed in silica gel particles is not the same as the maximum of the absorption spectrum of the compound in solution in alcohol. For precise and sensitive determinations the absorption maximum must be determined *in situ* on the plate. This is possible in the instruments with sources of variable wavelength or monochromators. For the instruments using filters, narrow-bandpass filters are available, and one should work as close as possible to the wavelength of maximal absorption.

In terms of the Kubelka-Munk theories as modified by Goldman and Goodall (see previous discussions), certain other conditions of the instrument must be considered.

1. Angular distribution of the incident light must be kept constant. This condition is met in most instruments presently available. In spectrodensitometers the light is almost completely

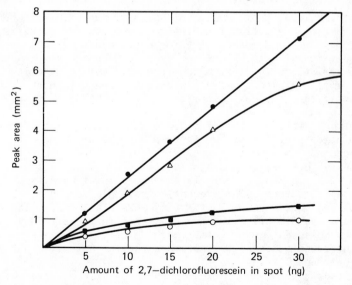

Figure 9.7 Regression lines resulting from exciting 2,7-dichloro-fluoroscein separated by TLC at different wavelengths. ● 310 nm, △ 300 nm, ■ 320 nm, O 290 nm

TABLE 9.5   EFFECT OF MEDIUM ON SPECTRAL CHARACTERISTICS
OF QUININE SULFATE

	Optical Maxima of Spectra (nm)		
Medium	Absorption	Excitation	Emission
0.1N H_2SO_4	250, 315, 350	340, 390	458, 466
Glycerol	315, 363	340, 390	450
Alcohol	280, 333	337	370
Silica Gel		315, 365	450

collimated.
    2. Monochromatic light is desirable. This condition is best
fulfilled in instruments equipped with monochromators. Otherwise
narrow-bandpass filters or wedge monochromators can be used.
    3. The aperture through which the incident light passes
should be small enough to illuminate a limited range of absorption.

Ideally the "flying spot" design of an instrument meets these conditions best. In addition to being rather expensive in terms of
system complexity and unnecessary for most chromatography applications, this scanning technique is for single-beam operation. Sub-
stantial variations in background signal are obtained in its use,
due to a small illuminated area, because of nonuniformity of the
layer itself, and various "impurities" caused by development of
the plate.
    For these reasons, the high noise levels and strongly chang-
ing absorption profiles of the layer along the plate make the
single-beam spot measurements very inaccurate. For double-beam
measurements there is little similarity between the measured zone
and the reference zone. There is an even noisier baseline. Thus,
for most applications using presently available sorbents, the flying spot technique may not justify its additional cost. Therefore,
double-beam one-dimensional scanning with a longer slit may be the
method of choice for quantitative methodology. Fortunately, most
of the instruments use slits, although they do not all provide
double-beam capabilities.
    For the range of absorption "seen" by a slit of given width,
under the approximation that the absorbance profile of the chromatographed band in the direction of development is an isosceles
triangle, Goldman and Goodall found (35) that the difference bet-

een the actual and measured absorbance could be measured. Fur-
hermore, for a known spot band width and the anticipated peak
ensity, a suitable scanning slit can usually be found. It is
ssumed that there is a negligible absorption gradient in the
ransverse direction along the band width. This is true only for
  short distance. However, even assuming that the areas where the
pot starts and ends do not contribute linearly, the peak area ob-
ained will be in most cases a reliable basis for quantitation.
n transmission the peak area (under the optical density curve)
s, within certain limits, proportional to the weight of the ab-
orbing substances, as expressed by the Kubelka-Munk formula.

   With the signal-to-noise ratio taken as

$$\frac{S}{N} = \frac{\text{peak height}}{\text{average noise}}$$

ower noise should be expected with wider slits. However, in
ouble-beam operations the noise remains fairly constant and the
ignal does not change appreciably with changing slit width, so
hat S/N may also be considered to be almost constant.

   Single-beam recordings (Fig. 9.8) show typical background
ariations of the plate and increased noise level in both trans-
ission and reflectance measurements. The figure shows the same
ackground smoothed out by a double-beam system. Although for
horter distances near the center of the plate the background
ariations and noise obtained by reflectance in single-beam opera-
ion seem to be acceptable, one has to realize that in reflectance
he peak response for the same spot is substantially lower than
nat in transmission mode, and consequently the S/N value (i.e.,
recision) should be expected to decrease.

   Figure 9.9 shows recordings of the same sample strip scanned
y transmission and reflectance. (The background noise due to
npurities in the plate was relatively high.) The peak signals
otained in transmission are about five times as high as the cor-
esponding signals in reflectance, the noise level being just
vice as high.

   Colorless UV-absorbing compounds are determined by reflec-
ance, or in the transmission mode by fluorescence quenching.
ith the latter method the average noise may be higher because of
neven phosphor distribution in some layers. However, this method
ives sufficiently large peak areas if the wavelength of maximum
osorption in the compound is close to the peak of wavelength
iving maximum fluorescence. Figure 9.10 shows an average noise
evel of a fluorescent plate (layer containing phosphor).

   In general, the desired linear relationship is not always
otained because the technique of spot application involves fac-
ors that affect linear response. For instance, varying areas per

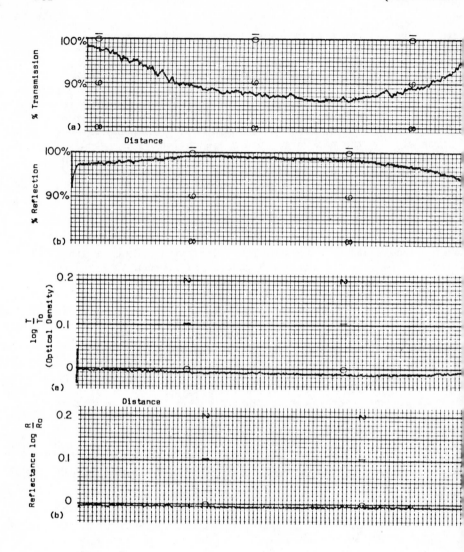

Figure 9.8. Upper curves: Scans of background in a TLC in the
single-beam mode. (a) Transmission; (b) reflection, both at
× = 420 nm, with slit width 0.7 mm. Lower curves: Scans of back
ground in the same TLC as above in the double-beam mode. Conditions are the same.

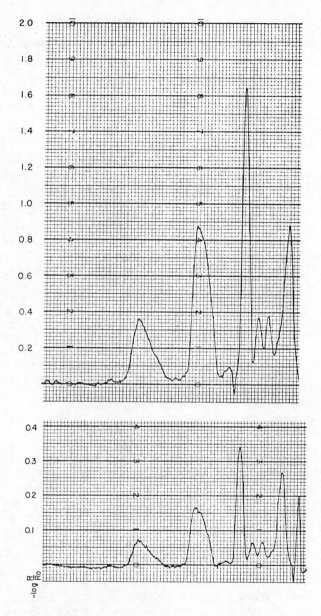

Figure 9.9. Comparison of double-beam (upper curve) transmission and double-beam reflectance (lower curve) of six components separated on a TLC.

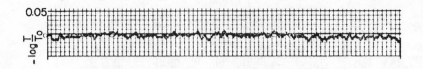

Figure 9.10.   Average noise in a TLC containing phosphor scanned
in the double-beam mode.

unit weight were obtained when samples were spotted with pipets
of different volumes.   Plates spotted with varying volumes of one
solution yielded linear relationships.   Attempts to repeat this
work with varying concentrations and constant volume resulted in
curved lines.

   According to Stahl and Jork (37), the peak area does not
change for the same weight spotted with varying volume if the
starting zone remains constant.   For increasing weight spotted
with constant volume and a constant starting zone diameter, they
obtained a linear relationship.   In other words, repetition of
the spotting technique under controlled conditions will yield re-
peatable results.

   Touchstone et al. (38) reported a linear response for steroids in the 0.02-10.0 μg range using a double-beam system.   Similar results were obtained by others.   The relative standard deviation caused by instrumental variations was found to be less
than 1%.   It seems reasonable to expect that if proper conditions
of measurement are maintained, a reproducibility of quantitative
determination better than ±2% can be achieved.

   There are a number of sources of error in any quantitative
determination involving thin layer chromatograms.   Systematic
errors are usually caused by application of an incorrect method.
These include unsuitable illumination and detection geometry of
the instruments, insufficient monochromaticity of light, and too
small a scanning aperture, as discussed before.   Operational errors may have the largest effect on the accuracy of a quantitative
determination.   A number of these sources will be discussed, since
generally the instrumental results only indicate poor technique in
the sample preparation and application, to say nothing of development of the chromatogram itself.

   Spotting errors are among the most significant sources of in-
accuracy.   The technique of spotting may affect the delivery of
volumes between 1 and 5 μl.   A constant starting zone size should
be maintained so that a given amount of substance spotted with
different volumes will yield identical peak areas, and different
amounts spotted with constant volume will exhibit a linear relationship in a plot of peak area of recordings of densitometer

cans versus weight. During movement of the solute through the
orbent layer, errors occur because of the lateral diffusion dur-
ng development, variations in the structure of different plates,
nd variations between tanks. Lateral diffusion depends on the
elative rates of diffusion of the solute and the mobile phase.
o avoid errors when substantially different amounts of solute
pplied next to each other diffuse out of the straight pathway,
lates can be scored into strips about 1 cm wide before spotting.
n this way two quite dissimilar samples can safely be developed
ide by side. The variation in the structure of different layers
eads to an incorrect extrapolation of the data obtained for a
nown quantity of standard to the quantity of the unknown. Since
he ratio between the quantity in the final spot and the quantity
potted varies for different layers, the use of a standard and a
est solution on each plate is recommended. In this connection,
actors such as the spot size before and after development and R_f
alues become important. Variations in results due to the use of
ifferent chambers are sometimes significant, and identical con-
itions should be maintained.

As discussed in Chapter 4 and pointed out by Touchstone et
l. (36), ideally, the sample or component to be quantitated
hould travel in the layer to a position of R_f between 0.3 and 0.7
or reproducible quantitative results. Shellard (39) also dis-
usses this variable in quantitative thin layer chromatography.
n R_f of 0.3-0.7 assures that there will be a more even distri-
ution of the solute in the zone on the chromatogram to which it
as traveled. Solute zones that are too concentrated militate
gainst linear calibration curves. Zones with low R_f travel are
ore concentrated than those with higher R_f values, and the sharp
arrow peaks obtained on recordings give areas not truly indica-
ive of the concentration in the zone. This effect is also seen
n the spectrometry of solutions. Solutions that are too con-
entrated do not follow the Beer-Lambert law. Consequently, if
ne considers that for the most part quantitative TLC is a micro
echnique, it is recommended that in most cases amounts of sample
ontaining no more than 1 μg of the solute be quantified.

The above, however, may also be construed as a general rule
n thin layer chromatography" "Start small." Overloaded chro-
atograms rarely present the separations that are possible when
he correct amount of sample is applied to the chromatogram. The
iterature abounds with reports of nonlinear calibration curves
hat may be due to use of overly concentrated samples.

A more recent development in densitometry is the appearance
f a densitometer using soft laser scanning (40,41). A number of
eatures in this instrument indicate a promising future for it.
he width of the beam is controlled internally rather than by a
lit. Thus, the beam is not diffracted as it would be if it pass-

ed through a slit. The beam's intensity can be adjusted so that
either TLC or gel layers can be scanned. With these features the
results of scans of high sensitivity show stable baselines. Quantitation will be based on interpolation from standard curves.

There is a new type of quantitative method for TLC based on
consecutive vaporization of the separated zones that requires
special but relatively low-cost instrumentation. Usually the products of the vaporization are either monitored by a vapor-phase
detector, as commonly used in gas chromatography, or collected
and analyzed by conventional means. The zones separated on a
standard flat chromatoplate cannot be vaporized easily. Consequently, the technique was originally carried out with layers
coated on rods (42), strips (43), or disks (44). More recently,
narrow tubes coated on the inner walls (45) were used, and instruments are now available for scanning tubular chromatograms.
An extension of this technique is the TAS system whereby the zones
are scraped from the layer and vaporized to be collected and
analyzed by conventional techniques (46-48). The vapors can also
be collected by condensation on the starting line of a TLC.

The scanner described by Mukherjee et al. (48,49) permits the
detection and quantitative analysis of organic substances in tubular thin layer chromatograms. The flow diagram of the scanner
equipped with a flame ionization detector is shown in Figure 9.11.
Samples to be analyzed are chromatographed in glass or quartz
tubes coated internally with a thin layer of silica gel or a mixture of silica gel and cupric oxide. After removal of the developing solvent, chromatograms are scanned. The substances in
the various chromatographic zones are vaporized consecutively,
either by pyrolysis or by combustion, and the resulting products
are monitored by a flame ionization detector or a thermal conductivity detector. Vaporization is accomplished by moving the
tubular thin layer chromatogram gradually through a ring-shaped
furnace while a carrier gas, nitrogen or helium, flowing through
the tube delivers the products of pyrolysis or combustion to the
detector. The products of pyrolysis are monitored by the flame
ionization detector, either directly (path B) or after their conversion to methane (path A). The carbon dioxide formed by *in
situ* combustion can be reduced to methane and monitored. If a
thermal conductivity detector is used (50), the products of combustion are passed through a water trap, and the carbon dioxide
in the effluent is monitored. With air or a mixture of nitrogen
and oxygen as carrier gas (51), the organic substances are vaporized mainly by combustion, from layers of either silica gel or
silica gel containing cupric oxide. In the recording of the
detector signals, both from a flame ionization detector and a
thermal conductivity detector, the various zones of the tubular
thin layer chromatogram appear as peaks whose areas correspond to

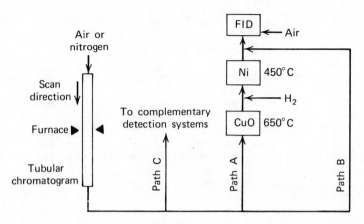

Figure 9.11.   Flow diagram of a scanner for tubular thin layer
chromatograms.   Path A, detection techniques PCRD, PCD, CD and
CRD; path B, detection technique PD; path C, complementary techniques, such as coupling with a gas chromatograph.

the total carbon present in the respective zones.

For radioassay of tubular thin layer chromatograms, a proportional flow counter is employed for monitoring radioactivity.
Low levels of radioactivity in various zones of a tubular thin
layer chromatogram are determined accurately when the scanner is
coupled with a trapping device.   The labeled compounds separated
in tubular thin layer chromatograms are quantitatively converted
to $^{14}CO_2$ and/or ^3H_2O, either by pyrolysis and subsequent combustion over cupric oxide or by combustion in $situ$ on absorbent
layers containing cupric oxide.   The products of combustion are
collected serially or continuously in Hyamine solution, and the
radioactivity of the fractions is measured by liquid scintillation
counting.

Both radioactivity and mass in various zones of a tubular
thin layer chromatogram can be monitored simultaneously by means
of a stream splitter.   Both radiochemical and chemical purity of
labeled compounds can then be assessed and their specific activities determined.

Mangold and Mukherjee (53) have summarized their experiences
with lipids using these techniques.   For quantitative use a
calibration curve must be set up for the compound to be assayed.
The compound, of course, must be separated on the tubular TLC,
vaporized, and if detected by a flame ionization detector, it can
be quantitated.   This is done on the basis of interpolation of the
curve by plotting areas of the corresponding peaks from the record-

er against the starting amounts. Essentially, the method represents an *in situ* quantitation of the TLC and can be very sensitive.

9.7.2   Calculation of the Results of Scans

As in any quantitative method based on interpretation of peaks
from the responses recorded from the detectors, there must be som
way to calculate the results. The recordings from densitometer
and radioisotope scans show peaks corresponding to the compounds
of interest. The use of present-day densitometers capable of reproducibility in scanning has made TLC a technique of high accuracy and reproducibility. Peak factors also are the basis of
quantitation in radiochromatogram scanning.

There are several possible sources of error in quantitating
the results of densitometer or radiochromatogram scannings.
These include:

1.   Sampling methods
2.   The separation variables
3.   Variation in detector responses
4.   Techniques of calibration and measuring

The discussions on sampling methodology in Chapter 3 provide the
beginning for good quantitative work. Production of reproducible
quantitative data is highly dependent on sampling techniques.
This involves proper preparation of the solutions for either the
densitometric or radiochromatogram scanning and the method of
transferring an aliquot to the plate. Good analytical technique
are important here as in any quantitative method. Precision wil
be determined primarily by these two factors.

Most errors in quantitation originate in the application of
sample to the layer. The techniques used to apply the sample wi
affect the shape of the spot after chromatogram development. The
process of the chromatography will determine whether there is ta
ing or overlap of the spots. High R_f values result in wide peaks
that are not amenable to reproducible results by scanning *in sit*

As recommended before, it is important for precise scanning
that the R_f of the components to be measured should be between
0.30 and 0.70. R_f values that are too high result in diffuse
spots, which give broad peaks that do not give true densitometri
results. Likewise, too low an R_f results in a very sharp, narro
peak in the recording that does not represent the true concentration in the zone scanned. Once again it must be pointed out that
the nonlinear calibration curves often seen in the literature are
generally due to overloading of the chromatographic system. Star
with 1 µg of material and work to the lower levels. Most TLC pro

cedures are capable of quantitation in the nanogram levels. Picogram levels are possible, especially with fluorometric procedures.

Another source of error is sample decomposition during the chromatography. This can be tested by developing chromatograms with standards of known serial dilutions. If a linear increase in peak area is not obtained as a function of concentration, sample concentration in the zone should be suspected. Errors due to tailing and overlap are often seen and can be eliminated by changing the mobile phase to obtain better separation.

The response of the detector in the instrument can be a source of error. Sensitivity, noise, and linearity are important in quantitative work. Furthermore, detector response is not the same for all wavelengths, nor do all substances show the same relative absorptive properties. The calibration must be carried out with the individual compound being measured.

Recorder tracings are generally used for evaluating quantitative results of densitometric and many radiochromatogram scans. The recorder should be properly adjusted. The methods used for calculating the peak-quantity relationship of the recordings are also subject to error.

There are a number of ways in which the area of the peaks of recordings of scans can be determined. Peak area may be determined with a planimeter, a device that computes the area of the peak as it mechanically traces the peak's outline. The precision and accuracy of planimetry depend on the skill of the operator. Precision can be improved with this technique by making repetitive tracings; however, this makes the time-consuming technique even more tedious.

A simple but very effective technique is to measure areas by the product of peak height, H, and the width, W, at one-half the peak height. This technique should only be used with symmetrical peaks or peaks that have similar shapes. The precision of this measurement can be improved by increasing recorder chart speed, so that the width of the peak can be determined more precisely.

The peak-area calibration procedure gives calibration based on area. Usually, the precision of this method is less influenced by changes in instrumental parameters. In addition, improved results are obtained with peaks whose shapes are not Gaussian. However, results from the peak-area quantitative method are more affected by neighboring peaks. Peak area should be used for better precision, and peak height for maximum freedom from possible interferences.

A simple method of quantitation involves the measurement of peak heights. Calibration is obtained by chromatographing serial dilutions of the compounds of interest and measuring the heights of the peaks on the recordings. Peak height is measured as a dis-

tance from the baseline to the peak maximum. Baseline drift is
compensated by interpolation of the baseline between the start and
finish of the peak. A standard calibration curve is obtained by
plotting the concentration of the compound of interest versus the
peak heights obtained. Unknowns are analyzed by chromatographing
the same size aliquot of an unknown solution and determining the
peak height of the materials for which the calibration has been
prepared. By interpolation on the standard curve, peak heights
can then be translated into the concentration of material in the
unknown sample.

The peak-height method should be used only when peak heights
change linearly with sample size. Serious errors will occur if
the peak-height method is employed when peaks are distorted or
the plate is overloaded. The peak height method is usually no
more precise then about ±5%, relative, since the results depend
greatly on the precision with which the sample has been applied.
Reproducibility is improved by using frequent calibrations to compensate for changes in instrumental parameters. The peak-height
method is valid only when the R_f values are reproducible. Low R_f
values give narrow peak widths on recordings of TLC scans. Likewise, zones with higher R_f values show broad peaks. Thus, if the
retention time in GC or the R_f in TLC is not reproduced within
narrow limits, the peak-height method is not valid. To overcome
this, the peak-area procedure should be used.

Peak-height measurements are more accurate than peak-area
measurements because peak heights are less interfered with by
neighboring, overlapping peaks. The peak-height method of quantitation also involves the least effort. It is often desirable to
initially determine the approximate concentrations of components
in a mixture by a simple peak-height method, before attempting to
set up a more rigorous high-precision procedure for routine use.

Another technique is to cut the peak out of the recorder
trace and weigh it. This method destroys the recorder trace. If
it is desired to retain the tracings, the recordings can be photo-
copied and the peak cut from the copy. The accuracy of this approach depends on the constancy of the weight of the paper and the
care used in cutting out the peak. By keeping the ratio of the
peak height to the width at half-height in the range of 1 to 10,
the accuracy of the cutting process can be improved. This method
can be used satisfactorily for irregularly shaped peaks.

Some recorders are equipped with ball-and-disk integrators
that automatically produce tracings that indicate the area of the
peak. The accuracy is still limited by the performance of the
recording potentiometer. The ball-and-disk integrator can often
be used with good accuracy for irregular shaped peaks.

The electronic integrators used in gas chromatography generally are also satisfactory for measuring peak areas in TLC scans

The electronic digital integrator automatically measures peak
areas and converts them into a numerical form, which is then printed out. Sophisticated versions of these devices also correct for
baseline drift.
    TLC peak areas from scans can also be conveniently measured
with the computer systems that have been developed for gas chromatography. There are many different types of computer systems
that can be used without change for TLC densitometry. With appropriate programming, most of these devices print out a complete
report, including names of the compounds, retention times, peak
areas, area correction factors, and the weight-percent of the various sample components. Snyder and Kirkland (54) state that the
maximum precision of the various integration methods is as given
in Table 9.6. Since the type of data generated by TLC densitometry is very similar to that found for gas chromatography, these
area precision values are representative for both techniques.

TABLE 9.6   AVERAGE PRECISION OF PEAK AREA MEASUREMENT TECHNIQUE
(NO INTERNAL STANDARD)

Method	Relative Precision, 1σ(%)
Planimeter	3
Triangulation	3
Cut and weigh	2
H × ½W	2
Ball-and-disk integrator	1
Electronic digital integrator	0.5
Computer	0.25

9.8   INTERNAL STANDARD TECHNIQUE

An internal standard will compensate for errors in the analytical
measurement. A known compound is added to the unknown mixture at
a fixed concentration. The added compound is also used as the
marker. The precision of the analysis does not depend on reproduc-
ing the size of the sample applied to the plate. Both peak-height
and peak-area ratio measurement can be used here. The compound/
internal standard peak-height (or peak-area) *ratio* is used for the
calibration.
    Peak-height ratio calibrations are constructed by chromato-

graphing aliquots of mixtures containing the compound of interest
in various concentrations together with a constant concentration
of the internal standard. The peak heights of the components of
interest are determined, and the compound/internal standard peak-
height ratios are plotted against concentration. This calibration plot is linear. Measurements by the peak-height ratio method
can be reproduced with a precision of 0.7-1%.

The peak-area ratio measurement technique is most precise,
since it compensates for changes in instrumental parameters and
variation in technique. The approach requires completely separated peaks in the chromatogram. The method of calibration is identical to the peak-height ratio method, except that measurements
are based on peak area.

The selection of the internal standard is critical for both
the peak-height and peak-area ratio methods. First, the internal
standard zone should be separated from the other zones in the mixture to be analyzed and located in a "vacant" spot in the chromatogram of the unknown. To determine the most feasible position
for the internal standard, an appropriate range of unknown samples
must first be chromatographically assessed. Optimum results are
obtained when the concentration of the internal standard is adjusted so that the peak-height or peak-area ratio is approximately unity. The internal standard must not be present in the original sample, and it should be a compound that is stable and not
reactive with sample components, sorbents, or mobile phase. If
possible, the internal standard should be commercially available
in high purity.

REFERENCES

1. W. A. Pons, Jr., A. F. Cuculla, S. L. Lee, J. A. Robertson,
    A. O. Franz, and L. A. Goldblatt, J. Assoc. Off. Anal. Chem.,
    49, 554 (1966).
2. L. Fishbein, Chromatography of Environmental Hazards, Vol. 9.
    Elsevier, New York, 1972, p. 398.
3. W. A. Pons, Jr., and L. A. Goldblatt, in Aflatoxins, L. A.
    Goldblatt, Ed., Academic Press, New York, 1964, p. 77.
4. C. A. Johnson, in Quantitative Paper and Thin Layer Chromatog-
    raphy, E. J. Shellard, Ed., Academic Press, New York, 1968,
    p. 105.
5. N. Oswald and H. Flück, Sci. Pharmacol., 32, 136 (1964).
6. J. C. Morrison and L. G. Chatten, J. Pharm. Sci., 53, 1205
    (1964).
7. H. Gänshirt and J. Polderman, J. Chromatogr., 16, 510 (1964).
8. G. Schierf and P. Word, J. Lipid Res., 6, 317 (1965).
9. M. Brenner, A. Niederwieser, and G. Pataki, in Thin Layer

Chromatography, E. Stahl, Ed., Springer-Verlag, Berlin, 1965, p. 391

10. H. J. Petrowitz, Mitt. Deut. Ges. Holzforsch., *48*, 57 (1962).
11. S. J. Purdy and E. J. Truter, Lab Practice, (*1964*), 500.
12. H. O. Bang, J. Chromatogr., *14*, 500 (1964).
13. H. H. Varon, H. A. Darnbold, M. Murphy, and J. Forsythe, Steroids, *9*, 507 (1967).
14. W. Heibrink, Fette, Seifen, Anstrichmittel, *66*, 569 (1964).
15. R. P. A. Sims and J. A. G. Larose, J. Am. Oil Chem. Soc., *39*, 232 (1962).
16. J. Haefelfinger, J. Chromatogr., *33*, 320 (1968).
17. W. E. Court, in *Quantitative Paper and Thin-Layer Chromatography*, E. J. Shellard, Ed., Academic Press, 1968, p. 37.
18. J. S. Matthews, A. C. Perede, and A. Aguilera, J. Chromatogr., *9*, 331 (1962).
19. G. Lehmann, H. G. Hahn, and P. Martinod, Z. Anal. Chem., *227*, 81 (1967).
20. J. Attal. S. M. Hendeles, J. A. Engels, and K. B. Eik-Nes, J. Chromatogr., *21*, 167 (1967).
21. J. C. Touchstone and M. F. Dobbins, J. Steroid Biochem., *6*, 1389 (1975).
22. V. Ganjam, C. Desjardins, and L. L. Ewing, Steroids, *16*, 1227 (1970).
23. B. A. Kottke, J. Wollenweber, and C. A. Cowen, J. Chromatogr., *21*, 429 (1966).
24. J. M. Brand, J. Chromatogr., *21*, 424 (1966).
25. H. Falk and K. Krummen, J. Chromatogr., *103*, 279 (1975).
26. J. M. Varga and F. F. Richards, Anal. Biochem., *53*, 397 (1973).
27. S. Udenfriend, S. Stein, P. Bohlen, and W. Dairman, Science, *18*, 871 (1972).
28. O. S. Privett and M. L. Black, J. Lipid Res., *2*, 37, (1961).
29. H. Jork, Deut. Apotheker Ztg., *102*, 1263 (1962).
30. O. B. Barrett, M. S. J. Dallas, and F. P. Padley, J. Am. Oil Chem. Soc., *40*, 580 (1963).
31. V. Novacek, Am. Lab., *2*(12), 129 (1969).
32. Chandrasekhar, Radiative Transfer, University Press, London, 1950.
33. P. Kubelka and F. Z. Munk, Tech. Physik, *12*, 593 (1931).
34. P. Kubelka, J. Opt. Soc. Am., *38*, 448 (1948).
35. J. Goldman and R. R. Goodall, J. Chromatogr., *32*, 24 (1968).
36. J. C. Touchstone, S. S. Levin, and T. Murawec, Anal. Chem., *43*, 858 (1971).
37. E. Stahl and H. Jork, Zeiss Inf., *68*, 52 (1968).
38. J. C. Touchstone, A. K. Balin, and P. Knapstein, Steroids, *11*, 115 (1969).

39. E. J. Shellard, in *Quantitative Paper and Thin Layer Chromatography*, E. J. Shellard, Ed., Academic Press, New York, 1969.
40. R. A. Zeineh, W. P. Nijm, and F. H. Al-Azzawi, Am. Lab., *7*(2), 51 (1975).
41. R. A. Zeineh and W. P. Nijm, Clin. Res., *22*, 428 (1974).
42. F. B. Padley, J. Chromatogr., *39*, 37 (1969).
43. J. J. Szakasitis, P. V. Peurifoy, and L. A. Woods, Anal. Chem., *42*, 351, (1970).
44. J. H. Van Dijk, Z. Anal. Chem., *236*, 326 (1968).
45. E. Hahti and I. Jaakonmaki, Ann. Med. Biol. Exp. Fennai, *47*, 175 (1969).
46. R. N. Rogers, Anal. Chem., *39*, 730 (1969).
47. E. Stahl, Z. Anal. Chem., *261*, 11 (1972).
48. K. D. Mukherjee, H. Soaans, and E. J. Haahti, J. Chromatogr. *61*, 317 (1971).
49. K. D. Mukherjee, H. Spaans, and E. J. Haahti, J. Chromatogr. Sci., *10*, 193 (1972).
50. E. Haahti, P. Vihko, J. Jaakonmaki, and R. S. Evans, J. Chromatogr. Sci., *8*, 370 (1970).
51. K. D. Mukherjee, J. Chromatogr., *96*, 242 (1974).
52. K. D. Mukherjee and H. K. Mangold, Ergebn. Exp. Med., *20'* (1974).
53. H. K. Mangold and K. D. Mukherjee, J. Chromatog. Sci., *13*, 398 (1975).
54. L. R. Snyder and J. J. Kirland, *Modern Liquid Chromatography* Wiley-Interscience, New York, 1974, p. 439.

CHAPTER 10
Radioactive Procedures

10.1 INTRODUCTION

Radioisotopes in TLC have wide application, particulally in metabolic studies, because of the great sensitivity that the methods can offer. Research in the metabolism of drugs and natural substances represents a broad field in which the localization of labeled substances in various tissues plays a great role. In order to identify and locate, as well as to quantitate these substances, the use of a radioisotope as a tracer for substances separated by TLC is important. Often, because of low detection limits, isotopic methodology is the means of choice, offering a number of methods by which these tagged materials may be detected or quantitated *in situ* after separation by TLC.

10.2 PURITY ASSESSMENT

Any discussion of the methodology in TLC of labeled compounds must define the limits within which one may work. Therefore, the purity and handling of isotopically labeled substances are a first consideration. When short-lived isotopes are used, one must be consistently aware that the material may not always be pure. Even with long-lived isotopes, such as ^{14}C, purity must be assessed, particularly if the samples have been stored for long times. Snyder and Piantadosi (1) feel that a labeled substance should always be assessed for purity immediately before use. A number of factors of contamination and degradation enter into this conclusion.

The high sensitivity of detection methods, especially in scintillation counting, would reveal very low levels of extraneous matter. Compounds not checked before use may show signs of instability after chromatography; even a contaminant of low radio activity can masquerade as a metabolite. Because of this frequent problem and because different compounds behave differently under the various conditions (that is, storage, type of solvent, type of sorbent, and their use in the separation methods), the entire process of the experiment must be evaluated in terms of each component that is to be assessed or quantitated.

Compounds of high specific activity can show autoradiolysis (2). Chemical changes can take place, especially in the case of soft beta emitters. Preparation of samples in dilute solutions can reduce the chance of this occurring. However, if the solvent is attacked and highly reactive, long-lived, excited species such as free radicals or ions are formed and further complications result. Water is a poor solvent in this regard, especially under aerobic conditions. Benzene is used by many companies that supply labeled substances to cut down these effects. When the compound is not soluble in benzene, alcohol or dioxane diluted with benzene are used. The samples should be stored at low temperatures to increase stability.

Chemical changes including those caused by air and light can also occur during the chromatographic procedures (3). This is true with either labeled or nonlabeled compounds. Snyder (4) indicated that the crucial stage is that when the solute is present on a dry sorbent and exposed to the atmosphere. Alumina and highly active silica gels are particularly troublesome in this respec Therefore, in quantitative work the results must be assessed as soon as the chromatogram is complete. In the authors' laboratory compounds have completely disappeared because of contact with the sorbent (silica gel) in the atmosphere after only 3 hr exposure.

The usual procedures for the handling of radioactive materials must be followed. In quantitative work particularly, contamination of the chromatogram or the sampling implements must be avoided, as pointed out in the chapter on sample delivery techniques. The delivery systems are of major concern in isotope work, since the sensitivity is so high. The use of calibrated disposable microcapillaries can save much time and effort in precise quantitative work using labeled compounds.

These basic beginnings are necessary to the success of quantitative procedures in TLC as well as other separation methods.

10.3   LOCATING THE SEPARATED SUBSTANCES

After the chromatograms have been developed, the separated solutes can be located in a number of ways. As discussed in the chapter on quantitative methods (Chapter 9), reference compounds can be used, or if the compound absorbs in the ultraviolet it can be visualized under the UV lamp. The detection methods described in Chapter 7 can also be used. The methodology of isotope detection in TLC is the subject of this chapter.

There are three main methods of detection when the separated solutes in the thin layer chromatogram are radioactive. These are:

1. Autoradiography *in situ*.
2. Liquid scintillation counting after elution, or in some cases after scraping from the plate without elution.
3. Direct chromatogram scanning.

10.4   AUTORADIOGRAPHY

Autoradiography, exposing the chromatogram to X-ray film, provides resolution comparable to the original chromatogram. The radioactivity level will determine the exposure time, which can vary from several hours or several days to weeks. The quantitative analysis is done by scanning the developed film in a densitometer in ways described in detail in the previous chapter on quantitation. The plate must be dried well to prevent formation of artifacts from traces of solvent left in the layer. Organic solvents can attack the film, since it is in contact with the layer, and produce spurious blackening.

Pseudoautography should be tested by chromatographing chemically identical but nonradioactive samples using the conditions required for the sample. If this possibility is not excluded, a thin plastic foil can be interposed between the chromatogram and the film to prevent *chemical* reduction of the photographic emulsion by the separated substances, solvents, or reagents. If the foil is not sufficiently thin, sensitivity of detection may be impaired for ^{14}C or ^{35}S. No foil can be used when the isotope is tritium. To prevent contamination of the film, handle the film as if it were contaminated. Do not touch it directly. In order to ensure the cohesion of dry layers and prevent their damage, they are often fixed by a protective spray such as clear Krylon or Neatan (see Chapter 8). Too thick a spray, however, may increase self-absorption, which is undesirable, especially if the spray is not applied in a perfectly uniform fashion.

Chromatographic spots are relatively large and diffusely lim-

ited. Thus it is desirable to have photographic films of maximum
sensitivity at the expense of fine grain. Single-coated film is
preferable for ^{14}C or ^{35}S, since blackening hardly takes place at
all in the distant emulsion, and background fogging is doubled in
two-emulsion layers. For autoradiography, place the X-ray film
on top of the TLC and cover it with a soft or elastic sheet (rubber or polyurethane foam, etc.) and a solid plate (wooden board,
metal or plastic plate), then weight or clamp it down. This operation must be done in darkness or under weak illumination (detailed by the instructions given by the film manufacturer). In
the case of penetrating radiation, the film can be left in its
black paper envelope and the work done with lights on.

   When a number of chromatograms of very weak beta emitters
are to be exposed for the same time, it is possible to repeat the
layers chromatogram-film-foam. Absorbent layers, such as lead
foil or 3 mm thick aluminum, must be interposed in the case of
^{32}P or ^{36}Cl. The sandwich is then wrapped in black paper (such
as the envelope in which films are purchased) or placed in an
opaque box and left for the time required. The time may vary in
practice between hours and months. To ensure correspondence between the chromatogram and its autoradiogram, markers such as
radioactive ink or heavy graphite pencil lines are useful. The
film is developed by the usual technique.

   In general, No-Screen Medical X-ray safety film can be used
for detecting the radioactivity. The emulsion is sensitive, and
the resolution suffices for TLC requirements. When tritium-la-
beled substances are to be recorded, Kodak NTB or Nuclear-Trak
emulsion films can be used.

   Radiation effects on a photographic emulsion result in the
appearance of a latent image that appears as silver grains after
development. Within a limited range of doses (radiation intensity
time), blackening is proportional to the concentration of the radioactive element in the spot. Above a certain limit, blackening
will not increase, and relative comparison of spot intensities becomes difficult. Overexposure must be avoided. In some parts of
the chromatogram, radioactivity may be relatively high, causing
merging of partially resolved spots, while other spots may be too
weak. A number of autoradiograms with different exposure times
should be prepared to assess the extent of these differences and
to determine optimal exposure times. If different isotopes are
present on one chromatogram, it is possible to differentiate between them by carrying out autoradiography with and without suitable filters. A film that is in contact with the chromatogram
can act as such a filter for another photographic plate.

   An adaptation of the autoradiographic procedure is to use
scintillation methods to improve the sensitivity. In the case of
tritium and similar soft beta emitters, the chromatogram can be

mpregnated with the scintillator, as reported by Wilson (5) (see
elow), or the scintillator (anthracene) may be added directly to
he layer before chromatography, as described by Lüthi and Waser
6). The chromatogram is then exposed to autoradiography. When
. liquid scintillator is used, be sure that no bubbles are trapped
nder the film. The substances to be chromatographed must be sol-
ble in the scintillator liquid used for the impregnation, or they
ay spread somewhat in the layer. Some TLC sorbents, especially
hose containing a phosphor, may act as scintillators by them-
elves (7).

The blackening of the film can be used to measure the sol-
te components as well as the radioactivity. However, to do this,
egression lines (radioactivity versus darkness) must be set up,
nd the range of linearity of the blackening must be assessed.
his means that the control of a number of chromatographic factors,
xposure times, and the amount of the radioactivity in the sepa-
ated zones must be considered.

The need for densitometry of autoradiograms and the long ex-
osure times make the method time consuming. The method is not
enerally applicable when a wide range of radioactivity or double-
abeled compounds are separated on the same layer. Consequently,
t is not widely used for quantitative purposes. A general auto-
adiographic method (8,9) that involves the apposition of a sheet
hromatogram to No-Screen X-ray film, is several thousandfold less
ensitive for ^{3}H than for ^{14}C. This difference is even greater if
ritium is compared with beta emitters of higher energy, like ^{32}P.
ecause of their low energy and short range, only a small pro-
ortion of the primary beta particles of tritium emerge from the
urface of the chromatogram (8). Further losses occur not only in
he gap between the chromatogram and the film, but also in the
ilm itsllf because the range of the ^{3}H beta particles in the film
s shorter than the average distance between the silver halide
rains.

Wilson, in 1958 (5), added a scintillator to the chromatogram
o convert the energy of the ^{3}H beta particles to light, which in
urn produced an image in the photographic emulsion. The sensi-
ivity reported for procedures of tritium autoradiography and
luorography that permit the subsequent recovery of the radio-
ctive compounds is 25-300 nCi/cm^2-day. If the chromatogram it-
elf is impregnated with an X-ray emulsion, a treatment that pre-
ludes the subsequent recovery of the compounds, the lower limit
f detection is reported to be 30 nCi/cm^2-day, according to
hamberlain *et al*. (8).

Randerath (9) made a systematic study of the factors that
nfluence the speed of tritium detection on chromatograms with
dded scintillator. The method makes possible the visualization
n thin layer chromatograms of 2-3 nCi ^{3}H per square centimeter

and of 0.05-0.06 nCi ^{14}C per square centimeter after an exposure
for 24 hr. Scintillation radioautography can increase the sensitivity of detection by a factor of 25-50.

10.4.1 Fluorography

For fluorography, a solution of the scintillator is either sprayed or poured over the dry chromatogram. A 7% (w/v) solution of
2,5-diphenyloxazole (PPO) (scintillation grade) in diethyl ether
is poured from a small beaker over the entire chromatographic
area as rapidly as possible. The scintillator is then distributed
evenly by tilting. This operation should take only 1-3 sec, depending on the size of the layer. The treated chromatogram is
then immediately brought into a vertical position and agitated
until the ether has evaporated completely. The volume of the PPO
solution used for this treatment is 35-40 μl/cm^2, that is, 14-16
ml for a 20×20 cm layer. All further operations are carried out
in a dark room under proper lighting conditions (7.5 W bulb,
Wratten 6B filter, distance 120 cm from the film). The layer is
placed in contact with the emulsion of Kodak RB-54 Royal Blue
Medical X-ray film (found to be the most sensitive) and kept between two glass plates in the dark in an insultated container over
dry ice (-78.5°) for the time necessary to visualize the labeled
compounds. The usual exposure time is 24 hr. The X-ray film is
subsequently developed using the manufacturer's prescribed procedure. Randerath also developed a procedure modified from the
early work of Muehler and Crabtree (10) for intensification of the
final image on the film as described below.

For preparing the four solutions needed, the chemicals specified are first dissolved in 400 ml distilled water and the solutions so obtained diluted to 500 ml with distilled water. The
following chemicals are required: *For solution* A, 5 ml 37% aqueous formaldehyde and 3 g Na$_2$CO$_3$; *for solution* B, 15 ml concentrated H$_2$SO$_4$ and 11.2 g K$_2$Cr$_2$O$_7$; *for solution* C, 1.9 g NaHSO$_3$, 7.5 g
hydroquinone, and 1.9 ml Kodak Photo-Flo 200 solution; *for solution* D, 11.2 g Na$_2$S$_2$O$_3$·5H$_2$O. For making the intensifier, 2 parts
solution C is added, with stirring, to 1 part solution B. While
the stirring is continued, 2 parts solution D and finally 1 part
solution B are added. The intensifier is made up fresh before use.
The stock solutions can be kept for several months. The technique
itself is as follows: After fixation the film is washed for 10
min in running water. It is allowed to harden for 5 min in solution A, washed for 5 min in running water, and briefly rinsed
under distilled water. The hardened film is then immersed in the
intensifier for 10-15 min, washed 10-15 min under running water,
and air dried. Only one film should be treated at a time; it is
important not to damage the emulsion mechanically in the areas to

be intensified.

For locating radioactive compounds on the original chromatograms, radioactive ink is applied as a marker to a few points on
the layer after chromatography. Following exposure, these spots
serve as reference points with which to line up the film and the
chromatogram. The compounds are marked by perforating the super-
imposed film around the darkened areas with a needle.

These procedures were developed for use with the "standard"
layers of 0.25 mm thickness. The results may be somewhat lower
if the layers are thicker. It should be noted that the spots on
the layer should not be so diffuse that a solute is of low concentration, since this would decrease the sensitivity of the autoradiography. A good rule to follow is to obtain migration of the
solutes of interest to the middle third of the chromatogram to
keep diffusion to a minimum.

The method will enable one to visualize 4-6 nCi ^{3}H per
square centimeter per day. With one-dimensional TLC (spot size
0.3 cm^2) 1.5-3 nCi ^{3}H can be located after a 24 hr exposure to
film. When expressed in terms of nCi/cm^2-day or dpm/cm^2-day, the
sensitivity is practically independent of the layer material.
Layers of cellulose, cellulose ion exchangers, polyamide, silica
gel, and silica gel-kieselguhr can be used. The green phosphor
of Merck F-254 layers does not increase the sensitivity of the
film detection method, regardless of whether or not PPO is present
and whether daylight or X-ray film is used. However, the radioactivity detectable in a given spot is lower on cellulose, PEI-
cellulose, or silica gel layers than on polyamide or silica gel-
kieselguhr layers because spots are more compact in the former
layers.

Prydz et al. (11) critically evaluated and improved the procedures of Randerath for a low-temperature solid scintillation
fluorography of TLC. The Eastman Chromagram sheet 6067 with silica gel is used. Anthracene is added by repeated dipping in a
benzene solution, or PPO is added with ether solution according
to the method of Randerath. The optimum amount of scintillator
was found from control measures in which the emitter luminescence
is measured by means of a sensitive photomultiplier. The dipping
is repeated until no further increase in beta-radioluminescence
(beta-RL) is observed. An additional scintillator amount would
have decreased the light output by self-absorption. This resulting decrease in the film blackening obtained was observed by
Randerath after extensive scintillator impregnation. The fastest
film available, Kodak RP X-Omat Estar Medical X-ray film, is used.
Two exposure temperatures, +20 and -78°C, are compared. Prydz et
al. used exposure periods of 5 hr, whereas Randerath used 24 hr.
Lüthi and Waser (6) exposed their films for longer periods of
time at room temperature, -30, or -70°C. The actual time for any

particular application should be determined by trail and error.

The optical densities obtained on the films can be evaluated
with a photoelectric densitometer. Curves showing the density as
a function of the exposure intensity (for a fixed exposure time)
can then be obtained.

In most investigations of the scintillation fluorography
technique, spots containing known amounts of radiotracer must be
used to evaluate the sensitivity. Randerath performs a complete
separation procedure and analyzes the results obtained for the
separated spots. To determine the spot activities, he used liquid
scintillation counting of the eluted spots. The solid scintillator luminescence output the various spots is measured directly
with a photomultiplier. To further test the behavior of the photographic film at low temperature and for low exposure intensities
he made some exposures using an electroluminescent lamp that gave,
approximately, a Poisson distribution of photons and for which the
light intensity could be varied stepwise with a series of neutral
density filters.

In thin layer chromatography the spots obtained are more concentrated than on paper. This is important for a low-level detection limit. The resulting number of grains per unit are, for
a first approximation, inversely proportional to the spot area.
However, the number of grains may increase when the area containing the activity is decreased, due to more effective production of
nucleation centers. What is important for their establishment is
the concentration of photons both in time and space.

The film can also be exposed with a mirror placed below the
chromatogram to increase the photon flux reaching the film. For
photographic plates with emulsion on both sides, the one not being
employed in the detection should preferably be covered watertight
during the development. Following fixation of both sides the fog
background will be reduced by 50%.

The optimum amount of scintillator should be used, and this
is more important for radiocarbon than for tritium. Because of
overlap between the absorption and emission spectra of the scintillators, they can exhibit self-absorption when applied in too
large an amount. Optimum amounts must be obtained by trial and
error. This is a well-known phenomenon for anthracene and has
been shown by Randerath to occur for PPO (9).

In connection with the temperature effect, it should be stated that different films show different behavior and have their own
particular optimum temperatures (9). For most scintillators, the
scintillator efficiencies vary only very slightly, and the cooling
of the pertinent solid scintillators should not be expected to produce more than about a 5-10% increase. On the other hand, frozen
benzene shows a drastic increase in efficiency upon cooling. The
use of a high-efficiency scintillator, although important, is not

in itself sufficient. One should also attempt to choose one whose
emission spectrum best fits the spectral sensitivity range of the
film material to be used. This can be determined from known characteristics of the scintillator and the specifications of the
film.

Adequate discrimination between the isotopes, usually ^{3}H and
^{14}C, has been difficult to obtain with those techniques that generally involve the application of multiple photographic emulsions
to the preparation. However, Gruenstein and Smith (12) described
a procedure for distinguishing autoradiographically between ^{3}H
and ^{14}C on a single TLC plate. Discrimination between the two
isotopes is good and the procedure is nondestructive, so that
samples may be recovered for further analysis. The method is
based on the technique of Randerath (9) in which the sensitivity
of ^{3}H detection was increased approximately 25-fold following addition of a scintillation fluor to the chromatogram. They observed that ^{14}C sensitivity is not enhanced by the scintillator
and suggested that it might be possible to distinguish between ^{3}H
and ^{14}C on the same plate by adjusting the ratio of ^{3}H/^{14}C radio-
activities and the exposure times so that only ^{14}C would be detected before, and only ^{3}H after, application of the scintillator.

The method is illustrated in work with amino acids separated
on cellulose layers. ^{3}H-Lysine and ^{14}C-leucine were applied on
an individual cellulose layer, and after chromatography an initial autoradiographic exposure was made for 24 hr at -78°C.
Following this, the plate was covered briefly with 7 ml of diethyl ether containing 7% (w/v) of 2,5-diphenyloxazole (PPO) and
dried in air at room temperature. The solution of PPO in ether
was applied rapidly with a pipet to the upper edge of the plate,
which was tilted approximately 10° from the horizontal. A second
autoradiographic film was then exposed in a manner identical to
the first, except that the exposure time was shortened to 4.5
hr. As described, this method should be applicable to all chromatograms in which the compounds being chromatographed are unreactive with the scintillator. If any of the compounds resolved
by the chromatogram are soluble in ether, care must be taken that
these compounds do not have time to diffuse or dissolve in the
ether.

Application of this technique to experimental situations requires consideration of certain additional points. If the material to be analyzed chromatographically contains several components, or if a single component is treated to yield several products, then, while the ration of ^{3}H to ^{14}C may remain quite constant for all products, the absolute amounts of radioactivity
would in some instances vary widely among the different chromatographic spots. In order to effectively analyze such a chromatogram, several exposures for different lengths of time, both

before and after treatment of the plate with scintillator, are
available. A wide range of absolute values of radioactivity can
be analyzed, provided that very faint spots do not overlap very
strong ones.

10.5 COUNTING BY LIQUID SCINTILLATION

Counting by liquid scintillation is probably the most commonly
used counting technique. Sections of the chromatogram containing
the sorbent are scraped off the plate. Then the solute is eluted
or the sorbent is transferred directly into the counting vial.
There are a number of ways that this can be done. Liquid scintillation is the most sensitive counting technique, such as will be
discussed in the next section, counting directly from the layer
is much less efficient than counting by the scintillation in solution.

The elution of the solute from the sorbent is subject to the
problems discussed in detail in chapter 9. The scintillation fluid, however, can be used as the eluting agent in a procedure in
which the sorbent is transferred directly to the counting vial.
Location of the zone to be scraped off for measurement of radioactivity is dependent on a reference marker. If the compounds are
visible or absorb in ultraviolet light they can be located. Resolution is limited by the width of the zone removed from the layer. To increase resolution, a large number of small sections can
be taken. The procedure of transferring the samples to the vials
is slow. Snyder and Kimble (13) described an apparatus that automatically scraps a predetermined width of zone into individual
counting vials. This apparatus is now available commercially.
The procedure is tedious, and when a large number of samples are
to be analyzed, it is subject to more sources of error. Generally,
however, most work with labeled compounds will involve assessment
of one, two, or three major components separated by TLC. As with
other quantitative methods, the standard must be carried through
the procedure so that regression lines and correction factors can
be determined for quantitative results.

Transverse zones on the plate are usually scraped off with a
spatula or blade. Suction devices run the risk of cross-contamination due to the passage of the powder through those parts of
the device that are used for removal of several zones. The chromatogram can be sectioned according to previous detection (physical, chemical, by *in situ* scanning or autoradiography) or at regular intervals, preferably intervals short enough to retain the
resolution of the chromatogram.

Individual sections or scrapings may then be eluted and the
eluate subjected to radioassay. This is seldom done, since part

of the radioactivity may remain uneluted and since elution involves a considerable amount of labor. In many cases, the sections or scrapings are counted directly. Counting of radioactive
materials after absorbing their solutions on cellulose or glass
fiber disks and evaporating is one of the standard techniques,
even outside the chromatographic field. It is possible to place
the section on a planchet and carry out counting, but liquid scintillation counting is more prevalent.

Snyder (14) showed that most detection reagents, including
iodine vapor, do not cause appreciable quenching of the scintillation. Silver nitrate quenches appreciably if it is used for
impregnation of a high concentration. Charring with sulfuric
acid causes very strong quenching in eluted zones and is not recommended in scintillation procedures.

10.6   ZONAL PROFILE SCANS

Snyder (15) claims that for quantitation, the liquid scintillation
radioassay of minute zones (zonal profile analysis) facilitated
by automatic scraping instruments provides the best resolution and
sensitivity for the detection of biochemical metabolites. Most
sorbents, visual indicators, and related materials that are used
in thin layer chromatography have no influence on the quantitative
aspects. Furthermore, low-level samples can be counted for long
periods of time so as to gain statistically good quantitative results. Zonal profile scans are especially helpful in exploring
labeling patterns of compounds in systems such as metabolic profiles in urine that are being studied for the first time, as the
scans can reveal peak areas of isotopic distribution that are not
associated with reference compounds. It is in this manner that
metabolic intermediates have been found. Details of the method
are given below.

After chromatography of the sample, the dried chromatoplate
is exposed to iodine vapor (or dichlorofluorescein) so that areas
of sufficient mass can be noted in relation to standard compounds
resolved on an adjacent lane. Next, the chromatogram is placed
in the scraping device that is capable of automatically collecting
small zones of the adsorbent layer along the entire chromatographic strip in counting vials used for liquid scintillation radioassay. A scintillation solution is then added, and the activity
in each vial is determined in a liquid scintillation spectrometer.
The vial is shaken for a predetermined time, and the gel is then
allowed to settle. Plotting of the data obtained from sequential
vials along the entire chromatographic lane results in a zonal
profile scan. The entire procedure of scraping, collection, dis-
pensing of the scintillation solution, plotting, and analysis of

the zones can be carried out automatically. Figure 10.1 shows
examples of zonal profile scans.

An important concern in the application of this procedure is
to ascertain that the total radioactivity in the labeled compound
applied to the chromatogram is accounted for by the integral of
all the zones collected from the origin through the solvent front
Quantitative recovery is possible only when self-absorption (due
to adsorption of the labeled component on adsorbent particles and
the glass surface of the vial) and other quenching phenomena are
absent. Snyder has found that relatively polar scintillation sol
utions containing a constant proportion of water permit the quantitative recovery of ^{14}C and tritium associated with both nonpola
and polar lipids on silica gel layers.

Dual isotope zonal profile scans show the problem of isotopi
cross-contamination caused by tailing of radioactivity from one
compound into an adjacent zone containing a different compound.
Techniques for the detection of radioactivity are extremely sensitive compared with the detection of mass. Therefore, the tailing of components is sometimes not obvious when the separated com
ponents on a chromatogram are made visible with spray reagents or
iodine vapor. On the other hand, zonal profile analysis of the
isotopic distribution along the lane will clearly show gross con-

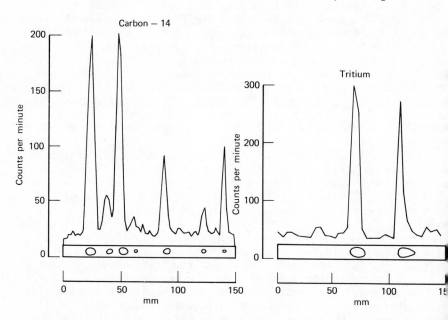

Figure 10.1. Examples of zonal profile scans. Courtesy of
Analabs, Inc.

...amination of unlabeled components in the system. If the tailing of a radioactive peak is not recognized, it will be mistakenly assumed that the radioactivity associated with a particular "spot" on a chromatogram is derived from that component made visible with the mass-detecting reagent. The shape of the radioactive zonal profile scan reveals maximal information about cross-contamination on the chromatogram.

The capabilities of the technique for obtaining zonal profile scans can be extended to the analysis of even smaller zones, for example, 1 mm, when extremely high resolution is required. This approach has already been shown to be effective in demonstrating the isotopic fractionation effect of tritium- and ^{14}C-labeled molecules. These experimental data also indicate the high resolving power of adsorption chromatography carried out on thin layers, even though under most circumstances such a high degree of resolution is undetectable.

If the radioactive material is soluble in the scintillation liquid and totally desorbed by it, no complications arise from the presence of sorbent in the vial. It is therefore worthwhile looking for a suitable scintillator mixture that would elute the material quantitatively. This is relatively easy with lipid-soluble substances. The addition of a polar organic solvent and a certain amount of water can even assist desorption, as reported by Turner (16). Experience has shown that the presence at the bottom of the vial of a white or transparent sorbent to which no radioactivity is attached exerts little, if any, influence on the counting efficiency. Black or highly colored material on the bottom of the vial reduces the reflecting surface and consequently lowers the energy of the pulses. Large amounts of sorbent on the bottom of the vial may reduce the energy of pulses generated by the external standard source, if the latter approaches from below the vial. Thus the plot of the ratio of efficiency to external standard channels is slightly shifted for ^{3}H and highly quenched ^{14}C samples; this shift is not apparent for slightly quenched ^{14}C.

Radioactivity that remains adsorbed on the sorbent may be counted with lower efficiency for several reasons. One is the less convenient geometry if the thin layer material lies on the bottom. The external standard cannot completely account for this effect, since it produces secondary beta particles and scintillations evenly in the whole volume of the liquid. The sample channel ratio method of correction, on the other hand, would be applicable if no self-absorption occurred.

Another way homogeneous dispersal of thin layer material throughout the whole vial can be ensured is by shaking with the thixotropic silica gel Cab-O-Sil. Particles of this preparation are so small that ^{14}C and ^{35}S, even if they were adsorbed, do not show any detectable self-absorption losses. Caution is necessary for ^{3}H.

Self-absorption in the fibers or grains of the sorbent must be taken into account in the case of ^{14}C, ^{35}S, and ^{3}H. Some of the beta particles emitted from the adsorbed substances in the direction of the sorbent grains are completely absorbed. This is not the case for ^{14}C if the silica gel grains are less than 10 μ in diameter. Normal silica gel for TLC purposes has grains 5-25 μ in diameter; thus, part of weak beta radiation may not be counted. This loss is not very great for ^{14}C. According to Snyder and Kimble (13), a ^{14}C-labeled compound that remains on the silic gel particles (10-25 μ) will give counting results approximately 8% less when it is homogeneously suspended in a scintillation gel than when it is in solution. For ^{3}H the respective figure is 25%.

Neither the external standard nor the sample channels ratio method will correct for the complete loss of scintillations. Internal standardization by the addition of a known amount of radio activity to the samples after they have first been counted may no be entirely satisfactory for sample self-absorption, even if the radioactivity did not differ chemically from that in the sample, since the added radioactive compound need not become absorbed to the same degree as that already retained by the sorbent. Careful calibration by graded amounts of substance subjected to chromatography, scraping, and counting in exactly the same way as the sample should eliminate the systematic error but can hardly be accurate, as the degree of sorption may not be reproducible and pipetting errors may affect the results. Quenching, including that by dissolved oxygen, would have to be corrected for individual samples in addition to the assessment of self-absorption losses.

If the labeled compound is completely dissolved, or if the sorbent containing beta emitters of sufficient energy (such as ^{32}P) is evenly distributed throughout the volume of the scintillating mixture, there is no serious objection to using external standardization, the most convenient way of correction. If the sorbent containing a sufficiently energetic emitter sediments, the sample channels ratio method is indicated, unless it is shown experimentally that the error committed by using external standardization is acceptable for the particular type of analysis. The sample channels ratio is less convenient, since it requires very long counting times in the case of low activities.

If the scintillation mixture does not dissolve all of a weak beta emitter, a reasonable compromise would be to use the externa standard (or sample) channels ratio and to derive, by internal standardization with samples of the particular substance subjecte to chromatography, the difference between the disintegration rates calculated from the known radioactivity added and from the external standard or channels ratio procedures. This difference, if

not very great, may then be used for correcting the results of
the analysis after the usual correction has been applied. The
chemical composition of the sample of known radioactivity used
for the calculation of self-absorption losses should, of course,
be as close as possible to that of the sample.

Lacko et al. (17) described a method that eliminates some of
the difficulties involved in scraping the sorbent from the TLC
plate. In a method limited to plastic supports or possibly the
plates with the foil backing, a punch is used to remove the zone
from the chromatogram rather than scraping the sorbent from the
support. In the method as described, TLC is carried out on plastic-based plates (Eastman Chromagram #6061), and the selected
zones are cut out by scissors or with a simple manual punch. The
punch is preferred when large spots are revealed by the scanning
procedure (iodine vapor), since scissor cutting results in some
flaking at the edges and thus causes loss of material. This
punch can be made by converting an inexpensive press (Manhattan
Supply Co., 171 Ames Street, Plainview, New York 11803) to produce uniform disks with a clear edge. The disks can then be transferred directly to vials for scintillation counting.

^{3}H-cholesterol (Mann/Schwarz) was found to be 95% radio-
chemically pure by TLC and was used without further purification.
The remaining 5% invariably stayed at the origin during chromatography and did not interfere with the radioassay. Recovery of
the isotope was tested by spotting identical amounts diluted with
plasma lipid extract on silica gel, and the disks were punched
out without subjecting the plate to chromatography or visualization. The same amount of ^{3}H-cholesterol was introduced directly
to vials for counting and taken as 100% for reference purposes.
All three gave essentially quantitative recovery. Toluene was
the most efficient solvent with the least amount of quenching. In
all scintillation fluids tested, the radioactivity was completely
extracted from the silica gel, for when the disks were removed
from the vials the counts did not change.

Since the extraction of lipids by the scintillation fluid is
such a crucial step in the counting process, another experiment
was carried out to test the counting efficiency and isotope recovery following TLC and visualization. ^{3}H-cholesterol was diluted with a lipid extract of human plasma prepared by the method
of Leffler et al. (18), and identical amounts were spotted on
plates. The recovery of radioactive samples can be substantially
accelerated without loss of accuracy when plastic-based plates are
used in combination with a manual punch. This technique is a sub-
stantial improvement over other procedures for the TLC of ^{3}H-
cholesterol. No extraction of the samples is required before the
radioactive counting, and self-absorption is avoided since the
material settles completely to the bottom of the scintillation

vials. The method should have general applicability in a wide
variety of systems provided that careful testing is carried out
to find the proper experimental conditions for each compound. See
Figure 10.2.

10.7 *IN SITU* EVALUATION OF TLC PLATES

Quantitative evaluation on thin layer chromatograms of radioactive
substances by direct scanning has not seen the attention that it
deserves in spite of the fact that as late as 1968 Snyder (14)
published a comparative table listing 11 scanners available from a
number of companies. The radioactivity may be measured directly
on the plate, with either a Geiger counter with a thin end window (19,20) or a gas flow counter (21,22). The gas flow counter
is more sensitive for the low-energy beta emitting isotopes, especially ^{3}H. In biochemical experiments in which tracers are used
to follow the course of metabolism, the technique of direct scanning of the radioactivity can be particularly useful.

10.7.1 Dot Printers and Spark Source Chambers

Pullan (23) introduced the spark chamber for rapid scanning of radioactivity on thin layer chromatograms. The procedure can also
be used for paper chromatograms.
    When beta particles flow between two high-voltage electrodes
placed above the chromatogram, a spark occurs that is recorded
photographically with a camera positioned in the top of the chamber. In the cross-wire spark chamber, the cathode (facing the

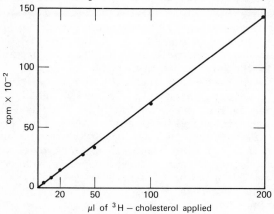

Figure 10.2. Relationship between amounts of isotope applied and
radioactive counts obtained.

underlying chromatogram) and the anode, which is about 2.5 mm
above it, are mutually perpendicular systems of parallel wires.
In the sturdier design of the coil spark chamber, the cathode is
in the form of several steel coils wound around thin anode wires
situated in their axes. Argon with methane or methylal is flushed through the electrode system. The resulting photograph re-
sembles an autoradiograph but is obtained with an exposure 1000
or more times shorter. The resolution is such that the edges of
two adjacent spots must be separated by more than 6 mm in the case
of ^{14}C and 3 mm in the case of ^{3}H. The pattern can be greatly
improved by "scanning," that is, by moving the electrode system
either manually or mechanically during the exposure. Decontamina-
tion of the cross-wire electrodes becomes a major problem unless
the chromatogram has been covered with a thin plastic sheet; this
is, of course, not feasible in the case of tritium. The advantages of the methodology here are that scanning times are short
and quantitative results can be obtained from measurements of peak
area (24). The method is nondestructive, and thus the chromatogram can be subjected to other analytical procedures.

More recently, Hesselbo (25) described a modified design of
Pullan's spark chamber. The work of Smith and co-workers (26)
has extended the use of this type of TLC scanning. In principle,
the spark chamber (as used by Smith) consists of a set of coiled
copper cathodes connected in parallel, each with a central stainless steel wire anode, these also connected in parallel. About
50-60 coils, each approximately 25 cm long, are cemented to a
plate of glass such that the overall useful area of the plate is
25×25 cm. Hence, the usual 20-cm-square TLC plate can be examined. Each cathode plus central anode constitutes an independent
condenser, and when any short is applied across the system a current will flow. As used here, a radiodisintegration will occur
at the position of any one of the separated radioactive substances.
This will be evidenced by a spark jumping the electrode gap, and
these sparks can be photographed to form a picture of the two-dimensional separation. The gap between the electrode plate and the
chromatogram lying on the base of the apparatus is a few millimeters, and the space is flushed with 100% methane in argon for
some moments before use. Photography is by means of a Polaroid
camera fixed in position at the top of a black chimney placed
above the electrode plate.

The apparatus is suitable for use with any isotope, function-
ing extremely well with ^{14}C and ^{32}P and less well, but quite adequately, with ^{3}H. Photographs can usually be obtained within 10
min, but exposures of up to 1 hr can yield good pictures, as can
such short exposures as 5 sec with chromatograms having zones of
high activity.

To locate the accurate position of the separated spots on the

original paper or layer, two procedures can be used. First, a radioactive grid is included with every chromatogram scanned to enable the picture to be realigned with the original. The grid is
a series of radioactive ink dots in two dimensions, one set of
dots running along the origin line and two sets of differently
spaced dots following the mobile-phase flow direction. The grid
and chromatogram are always placed in the apparatus in a standard
way so that mixed realigning cannot occur.

Second, the small Polaroid picture is placed face upwards and
covered with a sheet of glass or plastic film, and the spots are
carefully ringed. This sheet is then placed on an overhead pro-
jector, beamed onto a wall on which the chromatogram is pinned,
and enlarged until the grid spots from the beam become exactly
coincident with those of the original grid. Then the other spots
from the separated compounds must also be coincident with their
beamed spots, so these can be carefully encircled to mark their
positions permanently. Substances with very similar mobilities
can be readily delineated in this way.

It was originally hoped that quantitative data would also be
obtained via extensive computer hookup. Unfortunately, this is
not possible. Zones of very high activity act as spark sinks and
appear even hotter, and less active spots appear weaker than they
really are. Nevertheless, quantitation is readily achieved by
cutting out the respective areas, placing them in one of the usual
POPOP mixtures, and counting in a liquid scintillation counter;
differently sized zones have practically no effect on counting
background. In one sense, therefore, the visual picture can be
misleading, and it must always be remembered that the picture observed is entirely a qualitative one within very wide limits.

An example of the use of the apparatus follows. A thin layer
chromatogram on which nucleotides were separated was scanned. As
tritium decay particle energy is approximately one-tenth that of
the ^{14}C particle energy, the plastic sheet covering the spark
chamber had to be removed and the exposure time increased appreciably. Radioactivity of the different spots varied over a factor of
about 60 times, the lowest being 0.3% of the total of 3-4 μCi applied. Exposure of the whole chromatogram for 15 min at f22 showed only four major spots, whereas exposure for 60 at the smaller
aperture of f32 resulted in the appearance of at least five more
zones. When the first four major spots were masked out by six
thicknesses of Whatman No. 1 filter paper (1 mm thick) and re-
exposed for 1.5 hr with a camera aperture of f32, several more
spots appeared.

The masking-off procedure prevents the major spots from acting as electron sinks, thus allowing the most minor spots to be
visualized, including five not seen on the original autoradiograph,
which was obtained after a 4-day exposure. Hence, within a much

smaller time it is possible to take a number of photographs with the spark chamber, to mask off and/or excise the major radioactive spots, and to locate specifically the quantitatively minor spots or compounds present. The method has time-saving advantages when compared to radioautography.

10.7.2 Direct Scanning for Radioassay

There are several instruments available for direct scanning of a TLC for radioisotope detection of labeled compounds. There are applicable to both paper and thin layer chromatograms. They also provide nondestructible first analysis. However, they have probably the lowest sensitivity of the radiochemical methods discussed in this chapter, mainly because of the effects of layer thickness. Some of them are limited to narrow-width paper or TLC "strips." The more advanced instruments can handle the standard 20×20 cm TLC chromatograms. In contrast to the detection by "spark chamber" or autoradiography, which "photographs" the entire plate at once, scanners with a Geiger-Muller (GM) rate meter scan the plate as it moves at a constant speed under the window.

For one-dimensional chromatograms, the scanners might be the most convenient method of detection. The use of the film technique described previously is probably more widespread.

For scanning, the TLC is moved along under the detector at a constant speed and the count rate is recorded. After a suitable calibration, radioactivity present in a spot (or zone) can be calculated from the peak area on the record. GM tubes are available with or without a thin end window. An efficiency of 1.5% for ^{3}H may be attained. Tubes provided with a lateral aperture that serves as the slit through which the chromatogram is viewed are rather efficient (about 37%) for ^{14}C and ^{35}S.

If the chromatogram is in the form of sheets, rather than strips, it can be cut. There is the danger that in the case of oblique or irregular flow, individual chromatograms will be wrongly identified. Autoradiography or an informative GM scanning then becomes necessary; this may cause delay with low activities. In the case of glass-supported thin layer chromatograms, cutting is obviously inconvenient, if not impossible, but in some cases the layer can be removed. For thin chromatograms, one detector is generally used and the plates are scanned in a horizontal position. Resolution of adjacent peaks is given by the slit width, yet in the case of the Desaga scanner it was shown that, for a 2 mm nominal slit width, a detector situated 1.5 mm above the chromatogram yields peaks as broad as if the slit width were 12 mm. Decreasing the detector height above the plate improves the resolution.

A formula relating scan speed to percentage error and confidence level for a given length and radioactivity of a spot as well as the nominal slit width was given by Wood (27). Irregular oscillations of the recorder are due to statistical fluctuations in the disintegration rate. Noise can be reduced by increasing the time constant in the recorder circuit, but this broadens the peak and displaces its maximum toward later time intervals. Thus the lowest time constant compatible with acceptable noise level, the slowest scan speed compatible with the time available, the lowest detector height compatible with contamination risk, and a reasonably narrow slit will give optimal results.

In addition to the steady movement of the chromatogram through the counter, discontinuous movement is also possible in steps whose duration is given by either a preset time or a preset number of counts. Recording is either graphical or digital. In the case of a preselected number of counts, the program must ensure that impulse rates below a certain limit are skipped rapidly (low-background reject system) to prevent excessively long scanning times.

More recent improvements in instrumentation permit TLC scanning of as many as 12 individual lanes on a 20×20 plate. This can be done in either the X mode or an X-Y mode. In the first, after one lane is scanned, the plate is returned to the original position and then switched over to the next lane for a scan of that lane. This is repeated until all the lanes on the chromatogram are scanned or a preset number of lanes on the plate are covered. The detector does not count during the rapid right-to-left retrace. In the X-Y mode the plate moves in such a way that it zigzags left to right, right to left, down the Y direction. The Y increment depends on the width of the lane to be scanned. The instrument automatically shuts off when the detector reaches the end of the plate (see Figure 10.3).

The detector is usually a windowless flow-through Geiger chamber. A variety of slit widths, adjustable height, and windowless or Mylar windows or changeable diaphragms are also available. Also available are dot printers, which give two-dimensional radioactivity distribution. The dot intensity is proportional to the radioactivity present on the plate. The dot printer record is similar to that obtained by radioautography but is obtained in a short time.

The use of these scanners for any given application will be determined by the level of activity present in the chromatogram scanned. Scanning speed, height of the Geiger tube, slit widths, windowless or windowed, and the resolution of the zones on the chromatogram are variables that must be considered. The uniformity of the layer thickness is also important. Under proper conditions, as little as 100 dpm ^{14}C, 50 dpm ^{32}P, or 1000 dpm ^{3}H can be detected in a single spot. The counting yield, depending on

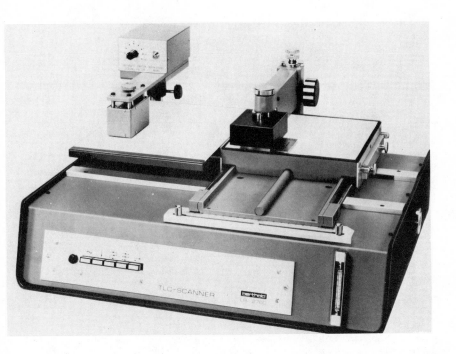

Figure 10.3.   Berthold instrument for scanning radioactivity on a
TLC plate.   Courtesy of Shandon Southern Instruments.

the layer thickness, is claimed to be as high as 15-30% for ^{14}C,
40-50% for ^{32}P, and 0.7-3% for ^{3}H.   Good reproducibility was attained in direct measurements of tritiated glutamate and 20-re-
duced desoxycorticosterone as shown by Wenzel and Brüchmüller (28)
Table 10.1 shows that quantitative recovery (14%) was obtained
when the same amount of radioactivity was applied to several chromatograms.   Recovery was good in spite of pipeting errors.   However, chromatography of a radioactive reference compound is recommended in assays of absolute activity because counting efficiency varies inversely with the thickness of the layer.
        The activity of a doubly labeled sample may be assayed di-
rectly, although the accuracy is reduced.   Wenzel (29) described
the principle of the method, which depends upon measuring collec-
tively the activity from ^{3}H and ^{14}C with the aid of a windowless
thin layer or 4π counter.   The activity from the ^{14}C is then measured separately using the attenuated efficiency of the end window (0.6 mg/cm^2) counter.   When a windowless scan is carried out,

TABLE 10.1   REPRODUCIBILITY IN DIRECT SCANNING OF
RADIOCHROMATOGRAMS

Radioactive substance	Activity (cpm) after 4 hr incubation*		
	0 mole-% NAD	5 mole-% NAD	500 mole-% NAD
Glutamate	37,900	13,900	13,600
20β-Hydroxysteroid	--	22,700	26,000
Total activity, cpm	37,900	36,600	35,200
%	104	100	96.5

* Activity of individual substances recovered at the indicated
  concentration of NAD (expressed as mole-% of the glutamate substrate).

both the ^{3}H- and ^{14}C-labeled zones on the chromatogram are detected.   When the end window is inserted into the GM tube, only the
^{14}C will be detected, thus giving a differentiated chromatogram
scanning as seen in Figure 10.4.

   In situ scanning of thin layer chromatograms was used for
detecting tritiated desoxycorticosterone and corticosterone after
incubation of the former with rat liver microsomes by Levin et al.
(30.   This has proven very useful in locating metabolites.   After
extraction of the medium with methylene chloride, the metabolites
were separated on silica gel GF thin layers using the mobile phase
chloroform:methanol:water (188:12:1).   The layer was scanned in
the fluorescence quenching mode at a wavelength of 250 nm on a
Schoeffel Model 3000 double-beam scanner.   Then the chromatogram
was scanned on a Berthold "Dunnschicht-Scanner" using the windowless GM tube.   As seen in Figure 10.5, a differentiation could be
made between the substrate, metabolites, and the extraneous unknown material separated from the extracts.   From interpolation of
the standard curves of the desoxycorticosterone (DOC) and corticosterone (B), the amounts of these substances could be determined
Furthermore, the conversion in the metabolism could be determined
as well as the specific activity.   This type of information is
becoming important as more research is being done in toxicology
and drug metabolism.   The direct scanning both with the densitometer and the radioscanner facilitates the location of compounds
of interest.   Thus, the in situ radiochromatogram scanner can be

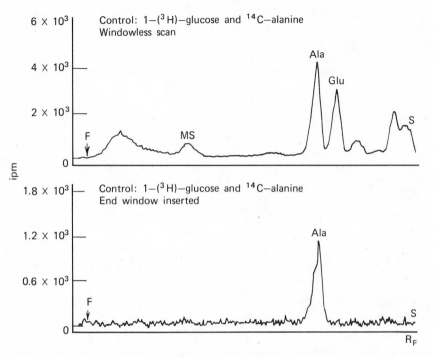

Figure 10.4. Use of end-window and windowless scans to differentiate ^{14}C and ^{3}H components separated by TLC. Ala: alamine; Glu: glucose. Courtesy of Shandon Southern Instruments.

a very useful tool in many areas.

Thin layer chromatography is a very rapid and effective method for purifying labeled compounds. In this respect it is also useful for assessing the presence or absence of impurities of breakdown products of the labeled substance. In most cases of organic or biochemical preparative radiochemistry it is advantageous to be able to scan for radioactivity without destroying the chromatogram. Therefore, chromatography with direct scanning of the separated radioactive compounds is the best way for quality control as well as purification of substrates. Burger (31) used the direct scanning method for assessing purity of a ^{14}C- and ^{3}H-labeled herbicide. The limit of detection for ^{14}C in thin layer plates was about 5×10^{-4} mCi per spot.

Although not as sensitive as scintillation counting, direct scanning of radioactive compounds separated on thin layers offers some advantages as pointed out in the previous discussions. With

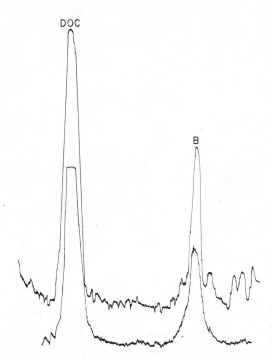

Figure 10.5.   Scan of a TLC plate representing a separation of corticosterone (B) and 11-desoxycorticosterone (DOC).   Lower curve: radioactive scan; upper curve: densitometer curve of fluorescence quenching scan.

newer and improved models of the radioscanners now available, the technique of direct scanning of thin layer separations should and will receive more attention.

REFERENCES

1.   F. Snyder and C. Piantadosi, Advan. Lipid Res., *4*, 253 (1966)
2.   R. S. Bayly and E. A. Evans, Storage and Stability of Compounds Labelled with Radioisotopes (Review 7), Radiochemical Center, Amersham, Netherlands, 1968.
3.   K. Macek, J. Chromatogr., *33*, 332 (1968).
4.   F. Snyder, Advan. Tracer Methodology, 4, 81 (1968).
5.   A. T. Wilson, Nature, *182*, 524 (1958).
6.   U. Lüthi and P. G. Waser, Nature, *205*, 1190 (1965).
7.   S. Prydz and K. S. Skammelsrud, J. Chromatogr., *32*, 732 (1968

.   J. Chamberlain, A. Hughes, A. W. Rogers, and G. H. Thomas, Nature, *201*, 774 (1964).

.   K. Randerath, Anal. Biochem., *34*, 188 (1970).

0.   L. E. Muehler and J. J. Crabtree, Phot. J., *86B*, 32 (1946).

1.   S. Prydz, T. B. Melstand, and J. F. Karen, Anal. Chem., *45*, 2106 (1973).

2.   E. Gruenstein and T. W. Smith, Anal. Biochem., *61*, 429 (1974).

3.   F. Snyder and H. Kimble, Anal. Biochem., *11*, 510 (1965).

4.   F. Snyder, Advan. Tracer Methodology, *4*, 81 (1968).

5.   F. Snyder, J. Chromatogr., *78*, 141 (1973).

6.   J. C. Turner, Intern. J. Appl. Radiation Isotopes, *19*, 557 (1968).

7.   A. G. Lacko, H. L. Rutenberg, and L. A. Soloff, Clin. Chim. Acta, *33*, 506 (1972).

8.   H. H. Leffler, F. W. Sunderman, and F. S. Sunderman, Jr., in *Lipids and Steroid Hormones in Clinical Medicine*, J. B. Lippincott Co., Philadelphia, 1960, p. 18.

9.   A. Breccia and F. S. Spalletti, Nature, *198*, 756 (1963).

0.   J. Rozenberg and M. Bolgar, Anal. Chem., *35*, 1559 (1963).

1.   P. E. Schulze and M. Wenzel, Angew. Chem., Intern. Ed. Engl., *1*, 580 (1962).

2.   P. Karlson, R. Mauer, and M. Wenzel, Z. Naturforsch., *18*, 219 (1963).

3.   B. R. Pullan, in *Quantitative Paper and Thin Layer Chromatography*, E. J. Shellard, Ed., Academic Press, New York, 1968.

4.   J. R. Ravenhill and A. T. James, J. Chromatogr., *26*, 89 (1967).

5.   T. Hesselbo in *Chromatographic and Electrophoretic Techniques*, Vol. 1, 3rd ed., I. Smith, Ed., Wiley, New York, 1969, p. 693.

6.   I. Smith, S. E. March, P. E. Mullen, and P. D. Mitchell, J. Chromatogr., *28*, 75 (1973).

7.   B. A. Wood, in *Quantitative Paper and Thin Layer Chromatography*, E. J. Shellard, Ed., Academic Press, New York, 1968, p. 107.

8.   M. Wenzel and M. Brüchmüller, Z. Naturforsch., *21b*, 1242 (1966).

9.   M. Wenzel, Naturwissenschaften, *52*, 129 (1965).

30. S. S. Levin, H. Schleger, D. Y. Cooper, and O. Rosenthal, Fed. Proc., *30*, 308 (1971).

31. T. Burger, J. Labelled Comp., *4*, 262 (1968).

CHAPTER 11

Reproducibility

11.1 INTRODUCTION

The field of chromatography has grown considerably since its introduction. However, even now there are many problems with the factors affecting reproducibility. Many practitioners of chromatography often look down on TLC as being a nonscientific, nonreproducible method. This, of course, is not true. Thin layer chromatography is often used as a qualitative method from which no reproducibility is generally expected. For example, the standard is chromatographed along with the sample, and the purpose is served. With the use of internal standards, TLC can be used for quantitative purposes with results as good and often better than other forms of chromatography. Often, also, TLC is the method of choice in many analytical procedures.

There are a number of reasons for poor reproducibility in TLC. Probably the most important one is inherent in the conditions under which the chromatogram is developed--the atmosphere in the developing chamber. This in essence calls for control of the environmental factors during development of the chromatogram, as well as during the sample application and handling. Reproducibility concerns R_f as well as quantitation.

11.2 EFFECT OF LAYER THICKNESS

Early in his work with TLC, Kirchner showed that by controlling conditions, the reproducibility of R_f values could be held to ±0.05 (1,2). The need for a uniform TLC plate is to some extent

301

satisfied by the modern commercial plates, which tend to have a well-controlled and reproducible thickness. The recent development of high performance TLC layers (3) having highly reproducible particle size as well as layer thickness will further advance the reproducibility of chromatographic results. Variation of R_f values with a variation in the thickness of the plate has been shown by other workers (4-7). Solvent flow is slowed in thick layers, and when the layer thickness varies on an individual plate the erratic effect on R_f value can become more pronounced. Pataki and Keller (6) show the effects of layer thickness on R_f values in a series of experiments.

Dallas (8) has pointed out, however, that especially with the more polar solvents, the heat of adsorption onto the layer can be appreciable. The dissipation of this heat should be less the thicker the layer, and one can expect the temperature to be greater in a thicker layer than in a thin layer. Therefore, if there is a decrease in R_f with temperature as normally found in adsorption chromatography, this heat effect can lead to an increased R_f with increased layer thickness.

Geiss et al. (9,10) showed that dry solvent vapor can displace water from a partially activated plate and thereby increase its activity. The displacement occurs more rapidly with a thin layer than with a thick layer. This can result in an increase of R_f value with layer thickness.

However, Jänchen (11) reports that when wedge layers were used with varying development techniques, it could be demonstrated that the influence of layer thickness on R_f in adsorption chromatography could be attributed to the effect of the vapor-phase situation. A comparison of a "saturated jar" and a "saturated sandwich" illustrated that certain development chambers commonly considered to feature saturated conditions are far from being saturated. Jänchen feels that the saturation and relationship between the layer and the vapor phase of the developing system are the controlling factors rather than layer thickness.

11.3 EFFECT OF MOISTURE ON THE SORBENT

It is well known that the water content of a sorbent affects the "activity" of that sorbent. The amount of moisture, directly related to activity, has a pronounced effect on R_f values as well as on resolution. Hermanek et al. (12) showed the effect of the activity of the sorbent on R_f values. R_f values increased with a decrease in activity. The effect of the activity of the sorbent due to varying degrees of water content is considerable. Care should be taken in handling of the chromatograms during all phases of the procedure, including storage. Herein lies one of the great

er faults of reproducibility between laboratories. Some environ-
ments have consistently higher humidity and pollutant levels than
others. Dallas (4) indicated that half of the total amount of
moisture taken up by silica gel at equilibrium was absorbed in
about 3 min. Even breathing on the plate during the application
of the sample could affect R_f values.

Geiss et al. (13) showed that alumina layers exposed to an
atmosphere of 65% humidity gained 2% moisture in 6 min and 3% in
20 min. They also indicated that the R_f of a substance will vary
as much as 300% between an atmosphere of 1% humidity and one of
80% relative humidity. Dallas (4) also investigated the effect of
humidity on R_f values.

Considerable changes in R_f values as a result of different
amounts of water vapor in the ambient atmosphere was seen in studies of the TLC of hypnotics by deZeeuw (14). With increasing
humidity (lower sorbent activity), a rise in R_f was noted.
DeZeeuw recommended the use of a constant-humidity room and indicated that reheating the plates after spotting to reactivate the
layer, is dangerous with respect to reproducibility. In other
work, he (15) pointed out that the vapors absorbed from the atmosphere greatly determine the separation and that partition chromatography rather than adsorption chromatography may be the true
factor in obtaining the separations seen.

11.4 EFFECT OF CHAMBER SATURATION

The effect of saturation of the developing chamber with the mobile
phases used for chromatography can be seen in the different R_f
values. If the chamber is not saturated, the solvent evaporates
from the layer, with evaporation increasing as the mobile phase
travels over the layer. This can change R_f values to the higher
ranges. Honegger (5), while investigating chamber saturation,
temperature, and activity of the sorbent (silica gel) found that
the activity of the layer influenced the separation to the greatest extent. Chamber saturation, particularly when the same mobile
phase is used for several chromatograms, can be a decided factor.
Each time a chromatogram is developed the saturation changes, and
the composition of the mobile phase changes due to the fact that
the sorbent absorbs one component of the mobile phase mixture pre-
ferentially. Each time the chamber is opened, the moisture content also change. Thus it is very difficult to control these factors in TLC as it is now practiced. Some of the other apparatus,
particularly the high-performance TLC system described by Starker
and Hampl (16), overcomes many of these shortcomings. The system
is closed. The solvent delivery is in a closed system. The chamber is small, and only small amounts of solvent are needed. The

apparatus use is limited to 5 cm plates, and circular chromatograms result.

11.5 EFFECT OF TEMPERATURE

Temperature has not been found to have too great an influence except in some cases (3,5). Increases in temperature cause an increase in R_f, probably due to increased evaporation from the layer with the resultant increase of mobile phase flowing over the
layer. Higher temperatures will also change the solubility factors of the solute in the mobile phase as well as the interactions
between the sorbent and mobile phase, although with small temperature changes the effects would be small.

11.6 EFFECT OF DEPTH OF DEVELOPING PHASE

The effect of distance from solvent level to the point of sample
application on the R_f will depend on the sorbent used, the compounds being separated, and the solvent used (16,17). Furukawa
(18) showed that when the developing solvent is a two-component
system used in separating mixtures containing compounds of both
high and low R_f's, the level of the origin will show an effect.
There is a separation of the components from the mobile phase in
the sorbent, and the compounds of higher R_f, which travel with the
less polar solvent, are not affected by the distance of the application zone above the solvent. The higher above the mobile phase
the sample is applied, the greater will be the effect of the separation of solvents in the sorbent. The sorbent will selectively
adsorb the solvent of higher polarity. The greater the total distance the mobile phase travels, the greater the effect it has on
the R_f values, since the solvent flow through the zone containing
the solute lasts a longer time.
    This is important when mobile phases whose components have
different polarities are used. For these reasons the depth of the
mobile phase in the developing chamber as well as the distance of
the starting line from the lower edge of the plate should be
standardized.

11.7 NATURE OF THE SORBENT

One of the most important factors in reproducibility between laboratories is the particle size differences that exist in the sorbents available from the different manufacturers. The quality and
activity can also vary from batch to batch from the same manufac-

turer. These generally affect the time of development but do not
greatly affect the R_f values. Geiss (19) showed that differences
of as much as 50% have been observed in the R_f values for separations using silica gel and alumina layers from several different
manufacturers.

Most manufacturers activate the TLC plates before packaging.
The heating varies from 105 to 110° for 30-60 min, mainly to
drive off the water and activate the sorbent. However, depending
on how the plates are stored, the conditions will vary from laboratory to laboratory. If plates are stored in desiccators or in
the presence of desiccants, care should be taken to protect the
plates during sample application. This can be done by placing
another glass plate on the sorbent layer above the sample applications zone.

The activity of alumina layers is controlled to a greater
extent by the temperature of activation. The greater the activity, the harder it will be to retain this property during storage
and handling of the plates. Reichel (20) has shown that in the
thin layer chromatography of insecticides by the normal procedure, R_f values are quite strongly affected by the humidity of
the laboratory atmosphere. It is time consuming and not very convenient to bring the layer to a precise degree of activity before
the development of the chromatogram. Methoxychlor, dieldrin, and
heptachlor were poorly resolved when the relative humidity was
14% but were completely resolved on alumina plates developed with
Skellysolve B when the relative humidity was 60%.

For this reason, Dallas (21) advocates the use of relative
R_f values, which are independent of the adsorbent activity over a
wide range. If this is the case, the conclusion is that the mean
energy of adsorption of a solute on the sorbent surface is constant over a wide range of activity.

More recently, Halpaap (22) investigated parameters of the
sorbent in specific systems used in thin layer chromatography. He
found that particle size and distribution are as important as the
primary characteristics such as the pore system and activity of
the layer since they control the flow characteristics of the developing solution within the interspaces of the particles and thus
within the pores.

Still another consideration in the nature of the sorbent is
the binder used in preparing the material before spreading it on
the plate. The binders used by different manufacturers are different. Some TLC plates have hard surfaces, while others seem to
be very friable, with layers that brush off readily. These differences in the nature of the layer have decided influences on the
R_f values obtained. However, at the same time the separating characteristics of the layer are also changing. This means that the
resolution on some plates will be better than on others, depending

on the nature of the compounds to be separated.

Standardization of particular sorbents selected from the
different types, with narrow pore distribution and particle size
distribution is prerequisite for the reproducibility of chromatographic processes. The present trend of TLC manufacturers is in
this direction.

A report of Ripphahn and Halpaap (3) has shown that control
of pore size can greatly improve the quality of separations and
quantitative results in TLC. The development of "silica gel 60"
(silica gel with pore size of 60 Å) is based on this concept. Th
particle size is also carefully controlled. Optimization of the
chromatographic procedure enabled adaptation of the method on a
reduced scale. Smaller plates can be used with shorter development times since efficiency of separation is much greater. This
means that precision in TLC methodology is now possible because
conditions can be more readily controlled.

11.8   EFFECT OF THE MEDIUM pH

Sorbents from different manufacturers apparently have differences
in pH. Although this may be due to different methods of preparation, it may also be due to the environment under which the TLC
plates are stored. Since TLC sorbents usually have high activity
they are prone to absorption of pollutants from the atmosphere.
Depending on the nature of the solutes being separated, such contamination can have a decided effect on the R_f of the solute as
well as on the resolution obtained.

Tichy (23) has indicated that silica gel G seems to act as a
buffer. Therefore, in its presence the autoreduction of molybdat
solutions must be suppressed by higher concentrations of acid in
the molybdate solutions used for detection or quantitation.

11.9   EFFECT OF SAMPLE SIZE

The concentration of the solution and the complexity of the solute applied to the layer can affect the R_f and the resolution.
This in turn will affect the reproducibility of the separation.
The effect will vary in different situations, and the individual
case must be evaluated by itself.

Application of too much sample tends to overload the chromatographic system, by exceeding the mobile phase capacity or the
linear capacity of the sorbent. This results in zones or spots
with tailing that are hard to pinpoint for measurement of the R_f
values and to evaluate for quantitation by spot size or direct
densitometry *in situ*. It is sometimes difficult to determine

whether the tail represents another component not resolved or is merely a trailing of the major component. Beyond this, the R_f value can increase or decrease with sample size depending on whether the isotherm is concave or convex.

The R_f values can also be affected by the ionic bonding between different components (solutes) in the sample. Attraction or repulsion of components of lower concentration by the component of higher concentration can result in changes of the R_f values of some components. Geiss (19) has shown the effect of applying overlarge quantities of solute to the TLC plate. The R_f's of sample components that are of too high a concentration appear to decrease in spite of the fact that the other components, not being of high concentration, move in a normal fashion. The R_f of a sample solute can be affected, that is, increased or decreased, by the proximity of other solutes as a result of attractive forces as mentioned above. For reference purposes, the R_f values of the pure solute chromatographed singly should therefore always be quoted.

11.10 EFFECT OF SOLVENT PARAMETERS

Depending on the mobile phase used, its reuse can affect the R_f value. With volatile mobile phases the more volatile component will continually decrease with use (23), or it can be diluted with volatile components of the mixture being separated. As a rule, one should change the mobile phase periodically, more often when more plates are being developed.

As the mobile phase advances through the sorbent layer, the ratio of mobile phase to stationary phase is not constant. It is smaller near the solvent front than some distance behind it. Dallas (4) found that if the plate is allowed to stand 15 min after the mobile phase first reaches the layer limit, the R_f values, although a little higher than if measured immediately, are more constant and reproducible. This is because the ratio of liquid to stationary phase has had more time to become constant over the length of the chromatogram.

The R_f values measured in this way are independent of the distance between the point of sample application and the solvent front if the mobile phase is a single pure solvent. With a mixed solvent, however, there often is a solvent gradient up the plate and R_f values are no longer independent of the distance the mobile phase has traveled. The distance, therefore, is often standardized at 10 or 15 cm.

The velocity of the mobile phase can affect the R_f because the rate of attainment of equilibrium is not instantaneous. In ascending development the angle the plate makes with the vertical can also affect R_f; a greater incline will result in increased solvent flow. The mobile phase velocity is also affected by pore and particle sizes.

Solvents of highest purity must be used for development in order to obtain accurate and reproducible R_f values. The presence of a small amount of impurity of different polarity can have a large effect on R_f values. Few investigators realize that, for example, chloroform usually has ethanol added as preservative. Distillation will remove this alcohol. Experimental records should note whether or not the chloroform contained preservative. Furthermore, autooxidation of aged chloroform forms polar reactants that grossly affect mobility of the solutes being separated.

When using mixed solvents a fresh mixture should be used for each chromatographic development to allow for changes in composition resulting from differential evaporation or adsorption during development or from chemical interaction between the solvent components.

REFERENCES

1. J. G. Kirchner, J. M. Miller, and G. J. Keller, Anal. Chem., *23*, 420 (1951).
2. J. M. Miller and J. G. Kirchner, Anal. Chem., *25*, 1107 (1953).
3. J. Ripphahn and H. Halpaap, J. Chromatogr., *112*, 81 (1975).
4. M. S. J. Dallas, J. Chromatogr., *17*, 267 (1965).
5. C. G. Honegger, Helv. Chim. Acta, *46*, 1772 (1963).
6. G. Pataki and M. Keller, Helv. Chim Acta, *46*, 1054 (1963).
7. L. Starka and R. Hampl, J. Chromatogr., *12*, 347 (1963).
8. M. S. J. Dallas, J. Chromatogr., *33*, 193 (1968).
9. F. Geiss, H. Schlitt, and A. Klose, Z. Anal. Chem., *213*, 132 (1965).
10. F. Geiss, H. Schlitt, and A. Klose, Z. Anal. Chem., *213*, 331 (1965).
11. D. Jänchen, J. Chromatogr., *33*, 195 (1968).
12. S. Hermanek, V. Schwarz, and Z. Cekan, Pharmazie, *16*, 566 (1961).
13. F. Geiss, H. Schlitt, R. J. Ritter, and W. M. Weimar, J. Chromatogr., *12*, 469 (1963).
14. R. A. deZeeuw, J. Chromatogr., *33*, 227 (1968).
15. R. A. deZeeuw, J. Chromatogr., *32*, 43 (1968).
16. L. Starker and R. Hampl, J. Chromatogr., *12*, 347 (1963).
17. E. J. Shellard, Lab. Pract., *13*, 290 (1964).
18. T. Furukawa, J. Sci. Hiroshima Univ. Serv., *A21*, 285 (1958).
19. F. Geiss, J. Chromatogr., *33*, 9 (1968).
20. W. L. Reichel, J. Chromatogr., *26*, 304 (1967).
21. M. S. J. Dallas, J. Chromatogr., *33*, 58 (1968).

22. H. Halpaap, J. Chromatogr., *78*, 77 (1973).
23. J. Tichy, J. Chromatogr., *78*, 89 (1973).

CHAPTER 12

Preparative Thin Layer Chromatography

12.1 INTRODUCTION

Preparative thin layer chromatography (PLC) may be defined as the thin layer chromatography of relatively large amounts of material to prepare and isolate quantities of separated substances for further work such as additional chromatography, infrared analysis, melting point determination, or synthesis.

Generally, all the procedures used with analytical TLC may be used with PLC, the major difference being that the thickness of the sorbent layer is at least twice that used in analytical work. Layer thicknesses range between 500 μ (0.5 mm) and 10000 μ (10 mm), but the commonly used thicknesses are 1000 μ (1.0 mm) and 2000 μ (2.0 mm). Commercially coated preparative layer plates are available from only a few suppliers in thicknesses of 500, 1000, 1500, and 2000 μ. Chapter 3 lists the plates available. Ritter and Meyer (1) chose 1-mm thick layers for the majority of their work because thicker layers did not consistently provide good resolution.

The resolving power of preparative plates is generally not as good as that of analytical layer plates, but PLC has a number of advantages compared with column chromatography, which is also commonly used for preparative separations. These include:

1. Smaller particle size (5-40 μ) of the PLC layer results

in sharper, distinct separations. However, small particle size column packings prepacked into tubes as "dry columns" are now available from a number of suppliers.

2. Conditions necessary for development of PLC may be experimentally determined beforehand by using rapid, analytical TLC.

3. Separated zones may be easily removed from a PLC plate and eluted.

12.2 PLATES AND LAYERS

Plate size for PLC is usually 20×20 cm or 20×40 cm, rather than the smaller sizes, because of the quantities normally dealt with. These two sizes are available commercially coated with preparative layers. Only Supelco, Inc. supplies a smaller PLC plate; it is 5×20 cm and coated with silica gel G or silica gel H with fluorescent indicator. For the preparative separation of extreme amounts, the sample may be applied to multiple plates, which can be developed concurrently in a large tank.

Many laboratories prefer to use commercially available (up to 2000 μ thick) PLC plates rather than making their own. Such use will avoid preparation problems such as the need for a special high-capacity spreader, the preparation time and mess, and the precautions that must be taken when handling, storing, and drying the thick layers. If layers thicker than 2000 μ are desired, it will be necessary for the researcher to follow certain precautions in preparing them.

The same quality sorbents used to coat analytical (250 μ) layer plates are not generally suitable for PLC plates. If a thick layer is made from a sorbent without a binding agent, the layer will be mechanically very unstable, rendering it unsuitable for use. When an analytical sorbent containing binder is used to overcome this problem, thick layers will most often pit, crack, or flake and thus usually be unacceptable for use.

To minimize these problems, sorbent manufacturers have formulate a series of sorbents especially for PLC. These are identified with the suffix letter P, as in silica gel PF-254. These sorbents are a mixture of different size particles to improve adhesion and do not contain calcium sulfate. They are recommended for making layers from 500 to 2000 μ thick.

To fill the need for sorbents that can be used for still thicker layers, another series of sorbents has been formulated. These are designated "P+CaSO$_4$" and are suitable for layers from 2 to 10 mm thick. Just as analytical sorbents are not recommended for making thick preparative layers, the preparative layer sorbents are not meant to be used for thinner analytical layers. The

preparative layers 500-2000 μ thick are used most often, and for
this reason plates precoated with layers this thick are available
commercially. These layers have the necessary higher capacity,
yet still offer a good degree of resolution. Layers greater than
2000 μ thick are used primarily for coarse separation or presepa-
ration, as they do not have the resolution of the thinner layers.

Affonso (2-4) prepared 1-5 mm thick plates of calcium sulfate
that were hard but limited in their separation capabilities because of the adsorption characteristics of the calcium sulfate.

Kirchner (5) developed a method for preparing layers 3-12 mm
thick with a stainless steel frame with internal wires for support
rather than a supporting plate. The layers were formed from
silica gel with 20% gypsum binder, and because there was no support plate, both sides of the layer were available for high-capacity sample application.

For convenience to the user, Table 12.1 lists the preparative
layer sorbents that are available commercially. It will be noted
that these are limited to the two most widely used sorbents, alumina and silica gel, and that all contain a fluorescent indicator.
A number of firms supply the various celluloses used for TLC (see
Table 2.2), and although it is not specifically mentioned in supplier literature, these powders should be suitable for preparative
layers if desired, although the sample capacity of such a cellulose layer would be much smaller than (as little as 1/50) that of
a comparable thickness silica gel layer. The major factor here is
the separation process taking place; partition on cellulose does
not have the capacity compared to adsorption on silica gel.

A variety of precoated preparative layer plates are commercially available. These are listed with the analytical layers in
Chapter 3. In addition to the selection of sorbents and the convenience that they offer, these plates have hard surfaces with
high mechanical stability. The hard surface has two advantages;
it is less subject to damage through handling and it allows the
sample to be applied through manual or mechanical streaking. It
is difficult to hand-streak a home-made preparative layer because
of the softness of the layer. Mechanical application is usually
used in this case.

Table 12.2 gives the procedures for preparing 2 mm thick preparative layers. Additional procedures may be found in Chapter
2. Storage racks for 20×40 cm preparative plates are available
from Brinkmann Instruments, as is a Plexiglas desiccator large
enough to store the plates. A wooden storage cabinet that will
hold 24 20×40 cm plates is also available.

TABLE 12.1 COMMERCIALLY AVAILABLE PREPARATIVE
LAYER SORBENTS (BRINKMANN)

Sorbent	UV Indicator	Layer Thickness, mm
Aluminum oxide 60* PF-254	254	2
Aluminum oxide 60 PF-254+366	254 and 366	2
Aluminum oxide 150† PF-254	254	
Aluminum oxide 150 PF-254+366	254 and 366	2
Silica gel 60* PF-254	254	2
Silica gel 60 PF-254+366	254 and 366	2
Silica gel 60 PF-254 with CaSO$_4$	254	10
Silica gel 60 PF-254 silanized RP-2++	254	2

* Previously designated type T; 60 indicates the pore diameter in angstroms.
† Previously designated type E; 150 indicates the pore diameter in angstroms.
++ Reversed-phase applications.

12.3 SAMPLE APPLICATION

Because the general goal of PLC is the separation of quantities of substances to be isolated, than normal larger amounts of sample are applied to the thick layer. A number of factors must be considered, including (1) the amount of sample to be applied; (2) the type of application technique to be used--whether streaking, spotting, or other; and (3) the method of sample application to be used--manual or automatic. Chapter 4 covers sample application in detail.

The maximum sample load for a 1000 μ thick silica gel layer (20×20 cm) is about 100 mg, less for cellulose and alumina.

In preparative work the majority of samples are applied by streaking, and few methods other than this have been reported. Honegger (6) and Meyer (7) cut V-shaped troughs into the layer to

TABLE 12.2*

INSTRUCTIONS FOR PREPARING PREPARATIVE TLC SORBENT LAYERS. Recommended slurry mixtures are based on the quantity needed to coat 20×20 cm plates with a 2 mm thick layer.

Sorbent	Slurry†	Drying & Activation	Comments
EM Aluminum oxide PF	300 g + 300 ml dist. H2O (shake 1 min). After 15 min standing, shake again before use.	Approx. 15 hr horizontal at ambient; then 3-4 hr at 110-120°C.	Separations usually accomplished by multiple development. After solvent reaches upper edge of plate, layer must be air-dried before second development. Use a solvent that moves sample only 1-2 cm during first run.
MN silica gel P + CaSO4	150 g + 260 ml dist. H2O	Approx. 1-2 days at ambient; than 3-4 hr at 110-120°C.	Separations usually accomplished by multiple development. After solvent reaches upper edge of plate, layer must be air-dried before second development. Use a solvent that moves sample only 1-2 cm during first run.
EM silica gel PF	150 g + 375 ml dist. H2O (shake 1 min).	Approx. 15 hr at ambient; then 3-4 hr at 110-120°C.	Plates can be dried with an infrared lamp or heater.
EM silica gel PF + CaSO4	800 g + 1600 ml dist. H2O (shake 1 min until slurry is uniform).	Approx. 1-2 days at ambient; then 3-4 hr at 110-120°C.	Glass plates should be cleaned only with water, not with organic solvents.

315

TABLE 12.2 (continued)

EM silica gel PF, silanized	150 g + 270 ml dist. H_2O/methanol (2:1). (see comments.)	Approx. 1-2 days at ambient; no activation.	Shake until slurry is homogeneous; then add additional 18 g sorbent. Shake again and do not use for at least 15 min.

* Courtesy Brinkmann Instruments.

+ Figures given for water or solvent addition are in all cases approximate. Slight variations may be necessary or desirable to suit local requirements.

load the solid sample into. The troughs of Honegger were 1-2 mm
wide and half as deep as the layer. This is a tedious, time-consuming process that must be done carefully to avoid cutting the
trough down to the plate, which would result in a discontinuous
mobile phase flow.

Narasimhulu et al. (8) used a somewhat similar technique
called "direct spot transfer." After samples were chromatographed
and visualized by UV quenching, if it was desirable to further
purify or chromatograph the separated zones, they were moistened
with one or two drops of water, and the zone, down to the glass,
was scraped off with a spatula. The scrapings were immediately
transferred to a new plate in an area on the origin that had been
cleared of sorbent. The cleared area had approximately half the
diameter of the area from which the scrapings came. Another drop
of water was added, and the scrapings were carefully pressed into
place with a spatula. After suitable drying the plate was developed.

These techniques should be generally applicable to the application of small amounts of solid sample to a TLC plate.

Lehman and Field (9) developed an apparatus for transferring
spot zones from one layer to another for rechromatography.

Connolly et al. (10) used a knife to cut two lines, approximately 3 mm apart, along the origin of the plate. Then they
carefully blew away any sorbent dust, leaving a 3-mm wide ridge.
With this method, sample solution is applied to the ridge with a
pipet or syringe. The sample solvent is allowed to evaporate, and
the lines on both sides of the ridge are filled with sorbent by
placing a sheet of aluminum foil with a long, 1-mm wide slit cut
in it over the knife cuts and using a spatula to press the sorbent
in.

The generally preferred technique for sample application in
PLC is automatic streaking. Sample spotting, although a simple,
convenient method, can become quite tedious if a large volume of
sample has to be applied manually. A spot produces a circular
starting zone, which, particularly if it contains a relatively
large amount of material, can interfere with the completely uniform advance of the mobile phase. Lipid and fatlike substances
are very resistant to development in polar (aqueous) mobile
phases, especially when they are very concentrated in a spot.

Manual streaking must be performed with care, as it can damage the layer, causing irregular development. It is often not
entirely uniform, with the result that the separation suffers.

Figure 12.1 shows the separation of a multicomponent sample
solution that was carefully streaked by hand compared to an identical sample that was automatically streaked. It would be difficult to isolate a desired substance from its neighbor without
contamination from the hand-applied sample because of the waviness

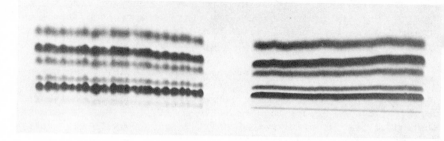

Figure 12.1. The separation of a multicomponent solution streaked by hand compared to automatic streaking. Approximately 150 µl
of a 5% solution of steroids was applied by hand-spotting on the
left and with the TLC Sample Streaker on the right. The plate
coated with Adsorbosil-5 was run in a solvent system of chlorofor
and acetone (98:2) and visualized by sulfuric-dichromate charring
From top to bottom: (1) cholesterol, (2) progesterone, (3) estrone, (4) dehydroepiandrosterone, and (5) testosterone. Courtes
Applied Science.

of the bands. This type of separation may be a hindrance in isolation because it is not always possible to visualize the entire
separation if the substances are not fluorescent or if chemical
visualization would change the nature of the substance. It is
sometimes necessary to assume a sharp, stright separation.

Considering these points, the best alternative is the streaking application with an automatic instrument. Other factors to
be considered are usefulness and cost. Sophisticated sample
streakers can cost anywhere from $400 up to $2800, and it would be
unwise to spend such money for an instrument that would seldom be
used. Laboratories that do a great deal of synthesis, or isolation of naturally occurring substances, may find it worthwhile to
invest in such an instrument. A number of commercial sample applicators and the characteristics of their operation are illustrated in Chapter 4.

Ideally, a sample streaker should have the following characteristics:

1. It must be able to apply the sample solution to the layer without damaging it. A damaged layer will impede the flow of
the mobile phase, resulting in an erratic and nonreproducible separation. This application is normally done without physical contact, which is important when fragile home-made layers are being
used.

2. The instrument should be capable of streaking the sample as a narrow line uniform in concentration and size over its entire length. This will allow for a sharp, distinct band separation upon development.

3. When quantitative evaluation of sample recovery is desired, it is necessary that the volume of the sample solution to be applied, and therefore the amount of sample solute, be accurately defined and reproduced. When applying the sample, it is important to avoid drop formation and evaporation at the tip of the applicator, because even a single drop may represent a relatively large percentage error. This is often considered unimportant in preparative work.

4. When applying larger solution volumes, with the higher solute concentrations normally encountered in preparative work, the mechanized sample streaking should preferably be from the same side and not back and forth. This will allow the application conditions to be as consistent as possible throughout the length of the streak. Drying time between applications should remain consistent to keep the band width of the applied sample compact.

5. The length of the streak to be applied should be adjustable to allow for plate size, amount of sample, and number of samples.

Quantum Industries manufactures the Linear-Q Preadsorbent Plate with a 1000 μ thick silica gel layer for PLC (see Chapter 3). This preadsorbent layer allows the application of up to 5 ml of sample, which does not have to be carefully spotted or streaked for effective separation. Such layers save valuable time in sample application.

12.4 DETECTION

The entire width of a plate should not be streaked, because there is a general tendency for the mobile phase to travel faster along the plate edge, which results in concave bands being formed upon development. When streaking a 20 cm wide plate, it is suggested that the streak be ended approximately 2-3 cm in from the left edge and 3 cm in from the right edge.

Before the plate is streaked, a suitably sized aliquot of the sample is spotted at the origin approximately 1.5 cm in from the right edge. After this aliquot is developed with the streak, it can be visualized by any technique desired without affecting the streak and thus serves to locate the separated components of the sample. This portion of the plate may be cut away from the rest of the plate, using a glass cutter for ease of handling during visualization. It is suggested that this be done if heating

is required for visualization. Heating may decompose the sample
components, and this procedure will allow visualization without
changing the composition of the sample. The large portion of the
plate may also be masked off from the portion to be visualized
by spraying by using a clean glass plate 20 cm wide.

If it is desired to apply known compound standards to the
plate to aid identification, keep them approximately 1 cm apart
and apply them next to the aliquot of the sample so that direct
comparison may be made. Another sample aliquot spotted at the
opposite end of the streak and visualized with the first one is
helpful in ascertaining exactly how the sample bands developed.
Figure 12.2 diagrams the general sample application procedure for
PLC.

Most precoated preparative layer plates are commonly available with 254 nm or 254+366 nm fluorescent indicators incorporated
into the layer. These plates should be used as much as possible
because they provide a good general nondestructive mode of detection, which is the most desirable mode for PLC. Other modes and
methods of detection suitable for preparative work are covered in
Chapter 7.

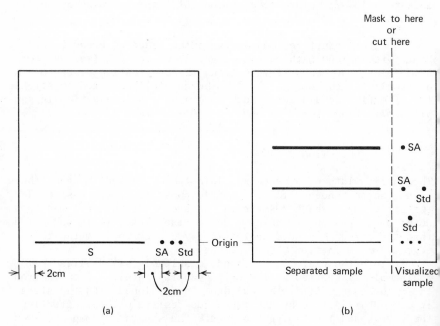

Figure 12.2. Sample application procedure for PLC. (a) Application. (b) Development and visualization. S = sample, saturated;
SA = sample aliquot; Std = known standard.

12.5 ELUTION OF DESIRED ZONES

After the preparative plate has been developed and the separated bands have been located, the desired substances may be removed from the plate in a number of different ways.

Camag has developed the Eluchrom automatic elution system, primarily for quantitative analysis work. This system will elute up to six zones that are no more than 2 cm in diameter, by passing a known volume of eluting solvent through the *in situ* layer and collecting it in a special vial. The eluent may then be transferred into a spectrophotometer or fluorimeter for quantitation. Vitek et al. (11) have completely described this apparatus.

When dealing with either long wide bands or small zones of separated substances, more manual methods must be used. These involve removing the layer containing the substance of interest completely from the plate, eluting the substance out of the sorbent with appropriate solvents, and then filtering and concentrating the solvent into a desired state for later use. Such procedures are illustrated in the literature (12-15).

The most common procedure is to simply outline the desired zone with a syringe needle or spatula and scrape off the layer down to the support onto a piece of glassine weighing paper with a spatula. The layer material is transferred into a small tube with a Teflon-lined screw cap. A few drops of distilled water are used to moisten the powder in order to inactivate it so that it will release the bound substance to be extracted.

In order to extract the desired substance for maximum recovery, a very polar solvent such as methanol, acetone, or water should be used. Often, however, water cannot be used because it would interfere with the subsequent operations to be carried out on the sample. Water is more difficult to handle than either methanol or acetone, which are preferred.

After the solvent is added to the tube containing the sample, the tube is capped and shaken for a given period of time, often only 5 min, by hand or mechanical shaker. The tube is centrifuged and the supernatant is pipeted carefully into another vessel for the subsequent operation, which is usually evaporation or concentration. At this point, it is generally recommended to extract the precipitated layer sorbent a second time following the same procedure.

To avoid scraping by hand with a spatula and the problems that could ensue such as loss by spillage or blowing, a number of different collectors for the layer material have been developed that are based on vacuum suction. Mottier and Potterat (16) were the first to publish the use of such a device. They applied vacuum to the constricted, cotton-containing end of a glass tube 8 mm in diameter, and sucked the layer up the open end, retaining

it on the cotton, which could then be extracted. Glass wool or a
200 mesh stainless steel screen supporting an asbestos mat may
also be used (14). Goldrick and Hirsch (15) used a sintered glass
disk to trap the sorbent.

A number of commercially made vacuum collectors are available. The simplest, shown in Figure 12.3, is manufactured by
Kontes and is available through a number of distributors. This
sample recovery tube has beveled, polished ends for use against
the layer and contains a sintered glass disk to retain the sorbent. After the layer material has been collected, the tube is
held vertically in a collection tube or flask and solvent is washed through the layer to elute the sample. It is generally good
practice to moisten the layer material in the recovery tube with
one or two micro drops of distilled water to inactivate the sorbent before the solvent is added.

Brinkmann makes a similar collector that employs a sintered
disk joined into a 14/20-neck round-bottomed flask (see Figure
12.4). The elution solvent may then be sucked directly through
the layer on the disk, and the eluted sample collected in the
flask. These collectors are limited in the amount of material
they may trap at any given time and were not intended to collect
an entire band off a thick-layer preparative plate. The general
apparatus must be scaled up to accommodate larger amounts of collected material.

Brinkmann manufacturers a "vacuum cleaner" type of zone collector that is designed to collect larger amounts of material.
The layer is sucked into a Soxhlet extraction thimble, which can
then be placed directly into a Soxhlet extraction apparatus for
efficient removal of the sample out of the sorbent material. Two
sizes of collector are available, a micro version that holds a
thimble of 2.5 ml capacity and a macro version that holds 65 ml
capacity thimbles. This apparatus was originally designed by
Ritter and Meyer (17).

When silica gel is eluted with methanol, soluble silicic acid
methyl ester may be formed and passed through the filtration medium
supporting the silica gel. To avoid contamination of this sort in
the final extract, evaporate the polar solvent to dryness and dis-

Figure 12.3. Sample recovery tube for TLC. Courtesy of Kontes.

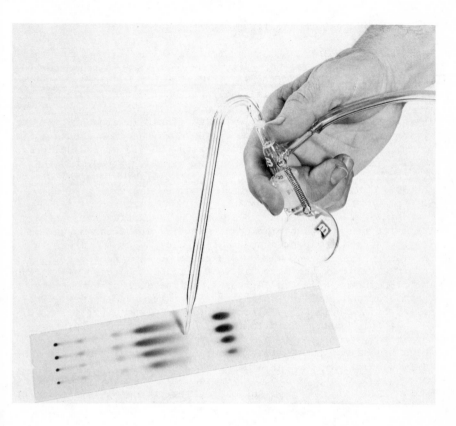

Figure 12.4. Spot collector with elution flask.   Courtesy Brinkmann Instruments.

solve the residue in a less-polar solvent such as benzene.   The
contamination will remain undissolved and suspended, and careful
filtration will produce a desired extract.

A useful piece of inexpensive equipment to have on hand when
eluting small areas from a TLC plate is a Swinney filter holder
for attachment to a standard syringe.   These are available in the
line of filtration apparatus carried by most laboratory supply
houses.   They are made of chromed brass and contain a removable,
perforated stainless steel disk for the support of a 13 mm diameter membrane filter.   The filtration side of the apparatus connects to a syringe, and the effluent end will retain a syringe
needle for delivery.   It is ideally suited for the filtration of
volumes of 1-5 ml of almost any biological fluid or any other liq-

uid that will not chemically attack the particular filter being
used.

This apparatus can be used to transfer a minimum amount of
the liquid being filtered. A syringe of a size suitable for the
volume of liquid being filtered is fitted with a needle. If the
volume of liquid is inconveniently excessive, the liquid may be
reduced in volume by evaporation, often with minimum or no heat,
which could cause decomposition. The liquid is drawn up into the
syringe, the syringe is turned upside down, and the needle is removed. The Swinney holder, fitted with the filter and delivery
needle, is then placed on the syringe, it is righted, and the
plunger is slowly depressed to filter the liquid into the desired
receiver. Nagel and Dittmer have reported their uses of the
Swinney holder (18).

Stutz et al. (19) have developed an apparatus for eluting
the large areas encountered on preparative layer chromatograms.
It is designed for continuous *in situ* elution of the areas in a
specially constructed chamber. Because of this feature, the procedure eliminates the mess and loss often encountered with scraping procedures and avoid carryover of fine sorbent particles into
the extraction (eluting) solvent. The interested reader is referred to the original publication for complete details.

If only small, analytical size samples are to be eluted, it
may be convenient to use plastic- or aluminum-backed layers,
which can be cut apart with scissors. The pieces containing the
desired zones may then be eluted (extracted) with a suitable solvent, as described earlier. The Eluquick apparatus, produced by
Camag, Inc., will do this automatically with up to 12 samples at
a time.

REFERENCES

1.   F. J. Ritter and G. M. Meyer, Nature, *193*, 941 (1962).
2.   A. Affonso, J. Chromatogr., *21*, 332 (1966).
3.   A. Affonso, J. Chromatogr., *22*, 452 (1966).
4.   A. Affonso, J. Chromatogr., *27*, 324 (1967).
5.   J. G. Kirchner, J. Chromatogr., *63*, 45 (1971).
6.   C. G. Honegger, Helv. Chim. Acta, *45*, 1409 (1962).
7.   H. Meyer, Chem. Abstr., *83*, 117990c (1975).
8.   S. Narasimhulu, I. Keswani, and G. L. Flickinger, Steroids,
      12, 1 (1968).
9.   M. C. Lehman and K. W. Field, J. Chem. Educ., *51*, 704 (1974).
10.   J. P. Connolly, P. J. Flanagan, R. O. Dorchai, and J. B.
      Thomson, J. Chromatogr., *15*, 105 (1964).
11.   R. K. Vitek, C. J. Seul, M. Baier, and E. Lau, Am. Lab.,
      6(2), 109-10, 112-16 (1974).

12. M. K. Seikel, M. A. Millet, and J. F. Saeman, J. Chromatogr., *15*, 115 (1964).
13. D. R. Gilmore and A. Cortes, J. Chromatogr., *21*, 148 (1966).
14. M. A. Millet, W. E. Moore, and J. F. Saeman, Anal. Chem., *36*, 491 (1964).
15. B. Goldrick and J. Hirsch, J. Lipid Res., *4*, 482 (1963).
16. M. Mottier and M. Potterat, Anal. Chim. Acta, *13*, 46 (1955).
17. F. J. Ritter and G. M. Meyer, Nature, *193*, 941 (1962).
18. J. H. Nagel and J. C. Dittmer, J. Chromatogr., *42*, 121 (1969).
19. M. H. Stutz, W. D. Ludemann, and S. Sags, Anal. Chem., *40*, 258 (1968).

CHAPTER 13

Special Techniques for Thin Layer Chromatography

13.1 INTRODUCTION

There are a number of special techniques used occasionally in TLC for specific purposes. Many require expensive and elaborate apparatus not found in the average chromatography laboratory.

13.2 DEVELOPING TECHNIQUES

13.2.1 Vapor-Programmed Development

Vapor-programmed (VP) development is a recent technique in which the effect of the vapor environment on resolution is used to optimize separation. Difficult separations can often be accomplished because optimum vapor conditions have been established over the entire plate. DeZeeuw has used this technique for sulfonamides and barbiturates (1,2) and local anesthetics (3).

For the convenience of the researcher, there are two commercially manufactured vapor-programming chambers. One, manufactured by Desaga and distributed by Brinkmann in the United States, is available in two sizes, one for 20×20 cm plates and one for 20×40 cm plates. This apparatus is shown in Figure 13.1 protected by a Plexiglas hood. This VP chamber consists of a metal block holding 21 troughs (each 20×1×2 cm or 40×1×2 cm), a solvent reservoir, and six sets of Teflon spacers (0.3, 0.5, 0.8, 1.0, 1.2, 1.5 mm thick).

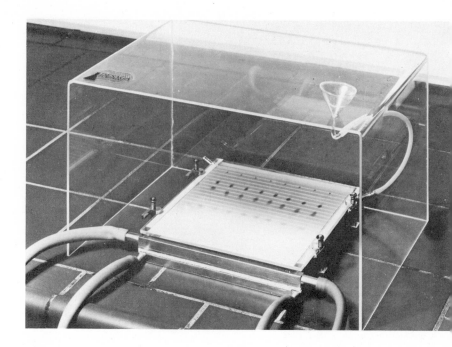

Figure 13.1.   Vapor-programming chamber.   Courtesy of Brinkmann
Instruments.

Using one set of spacers, the plate with the samples applied is
mounted with the layer side down toward the troughs, which have
been filled with suitable liquids.   The mobile phase is fed to the
TLC plate from a reservoir through a paper wick.   The different
vapors that have been absorbed by the dry layer before development
dissolve in the migrating mobile phase during development.   This
results in a controlled mobile-phase gradient, which is specifica
ly a reversed gradient.
     In conventional gradient techniques the solvent with the
lowest polarity is found in the solvent front; the opposite is the
case with the vapor-programmed technique.   Because the upper part
of the layer contains the more polar solvent, the faster migrating
compounds advance more rapidly and are resolved from the slower
moving compounds in the lower portion of the plate.
     DeZeeuw (2), using this technique, allowed the plate to re-
main in the chamber in contact with the solvent vapors for 10 min
prior to adding the mobile phase to the reservoir and starting
that development.   He states that "the actual advantage of VP-TLC

s that close lying spots can be pulled apart and then guided to
 position in the chromatogram which is not yet occupied by an-
ther compound. Accordingly, the entire plate length can be uti-
ized to cover the spread of the spots. Furthermore, due to the
act that any desired vapor composition can be selected and ap-
lied to any point of the plate, the analyst can select the opti-
al conditions for a particularly difficult separation of two com-
ounds, without disturbing the separation already obtained of the
ther components present in the sample . . . We indeed postulated
hat the increase in resolution which can be obtained in unsatu-
ated chambers could possibly be due to a concentration gradient
f vapor in the dry adsorbent. In order to be able to control
uch a gradient and/or to improve the gradient, we thus developed
he vapor-programming chamber."

   Geiss and Schlitt (4) developed another design of vapor-pro-
ramming chamber called the Vario-KS chamber, which is manufac-
ured and distributed by Camag, Inc. This chamber is shown in
igure 13.2. It is supplied with three different troughs that are
eferred to as conditioning trays: one with 5 divisions, one with
0 divisions, and one with 25 square divisions.

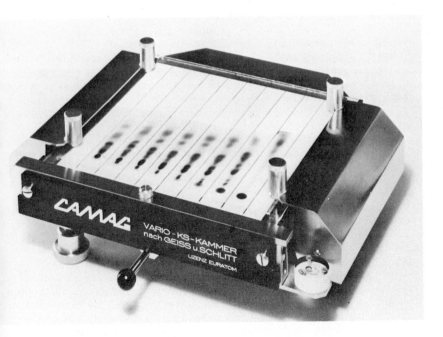

Figure 13.2. Vario-KS chamber for vapor programming TLC plates.
Courtesy of Camag, Inc.

The mobile-phase tanks are available both without divisions and with five equal divisions. Sandroni and Schlitt (5) reported the application of the latter for the concurrent development of a 20×20 cm plate with up to five different mobile phases, which are delivered to the plate with five paper wicks. Because of the dif ferent flow rates of the different mobile phases through the laye some sorbent must be scraped from the plate so that the layer con sists of five sections each 16.5×2.5 cm. To avoid mixing of the vapor phases between sections, a sandwich slide made of six 1 mm thick polypropylene barriers glued together is inserted between the sections before development.

Figure 13.3 shows three applications for the vapor-program- ming chamber. (a) Preloading with five solvents for the rapid selection of the most suitable separation system. A dye mixture was developed with pure ethyl acetate on silica gel after pre- loading the layer with five different solvents that were less po- lar than or equally polar to the ethyl acetate mobile phase. (Left to right: ethyl acetate, benzene, carbon tetrachloride, cyclohexane, n-hexane.) (b) Effect of relative humidity. Pre- loading with water-sulfuric acid mixtures in the conditioning trough resulted in eight humidity zones for the determination of the effect of relative humidity on separation. Shown is a dye mixture on silica gel. Three sequence inversions of the compo- nents are evident. Humidities, left to right, are 72%, 65%, 47%, 42%, 32%, 18%, 14%, and 9%. (c) Simultaneous development. A dye sample was developed in five different mobile phases simulta- neously to aid in the choice of the best phase for the separation Left to right, these were cyclohexane-benzene 1:1; benzene; di- chloroethane-benzene 1:1; dichloroethane; and diethyl ether-di- chloroethane 1:1.

The Vario-KS chamber is also available with a heating acces- sory for continuous development chromatography.

13.2.2 Radial Chromatography

In radial (circular) chromatography, the sample substance to be separated is spotted in the middle of a horizontal plate. The mobile phase for development is supplied to the center of the sam- ple zone by means of a wick or micro dropping pipet. The sepa- rated zones develop as concentric rings.

DeThomas and Pascual (6) have designed a simple apparatus for demonstrating radial chromatography, and Rachinskii (7) has con- sidered the basic theoretical principles involved in this action. This form of TLC is not as versatile or as convenient to use as normal development methods. For example, comparison with stan- dard substances, elution of separated zones, and quantitation are more difficult to perform than with conventional methods. Because

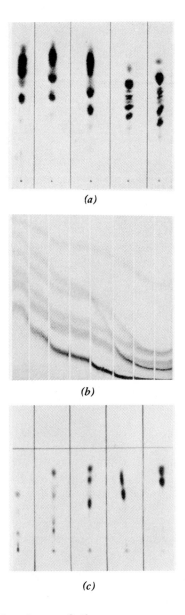

(a)

(b)

(c)

Figure 13.3. Applications of the vapor-programming chamber. (a) Choosing the most suitable vapor separation system. (b) Effect of relative humidity. (c) Choosing the best mobile phase for a separation. See text for details. Courtesy of Camag, Inc.

of these factors, radial chromatography is seldom used.

Schleicher and Schuell Co. is marketing a device called the SelectaSol Solvent Selector System, which employs circular devel opment to aid in the selection of a suitable mobile phase that can then be used for normal development. A template is supplied so that the sample spots may be applied in correct positions on the plate. The chamber base has 16 small wells in which up to 16 different mobile phases may be tested. A wicking plate contains wicks that immerse in these wells as it is sealed against the bas with a gasket. The spotted 20×20 cm plate contacts the wicks aid ed by the proper spacing of a gasket. The entire assembly is screwed together, and development is allowed to proceed horizon-tally.

Camag is offering an apparatus called a U-Chamber for fast screening circular chromatography. Separations normally require only 1-4 min, using the 5×5 cm High Performance TLC plates (HPTLC manufactured by E. Merck that are available from Camag and EM Laboratories.

In operation, the spotted HPTCL plate is positioned by a plate holder ring with its layer facing downward on the U-Chamber body. Mobile phase for development is fed to the center of the plate by a platinum-iridium capillary. Constant flow of the mo-bile phase is achieved by a motor-driven 250 μl syringe. Normal development requires this volume of mobile phase or less.

Vapor phase, made up externally, may be passed through the chamber and out a center bore before, during, and after develop-ment. The name of the apparatus is derived from the mode of operation, with everything occurring underneath.

13.2.3 Hot Plate Chromatography

According to Turina *et al.* (8), warming a chromatographic plate during the TLC process causes the solvents to evaporate from the plate, and the chromatograms thus obtained show two advantages: (1) better resolution of the spots; and (2) the possibility of detection of trace components that are undetectable under usual conditions. These effects, which are obtained under special con-ditions, can be explained by the discontinuous countercurrent mod el of the chromatographic process (9).

Turina and Jamnicki (10) consider the theoretical mathemati-cal principles of hot plate chromatography, which is seldom used in this form because it has little advantage for the separation o many compound types.

13.2.4 Programmed Multiple Development

Programmed multiple development (PMD) is an automated form of mul tiple development for TLC that was first reported in 1973 by Perry, Haag, and Glunz (11). In PMD, the thin layer plate is automatically cycled through a preset number of developments. In each succeeding development, the mobile phase is allowed to advance further. After each development, controlled evaporation by heat or inert gas causes the solvent front to recede, usually to or beyond the point of the initial zone of sample migration. After the last desired cycle, continued controlled evaporation prevents further development. PMD exploits the reconcentration that occurs each time the solvent front traverses the sample zone. As a result, spot braodening during development is counteracted. PMD is claimed to have advantages over conventional TLC in plate efficiency, speed, and sensitivity. These vary according to the operational parameters chosen.

An apparatus for PMD commercially available from Regis Chemical Co. is illustrated in Figure 13.4. It consists of two units, a programmer (left) and a developer (right). The programmer is a computerized controller that directs predetermined PMD parameters to be carried out by the developer. The developer, as shown in Figure 13.5, consists of a solvent trough, a sandwich chamber for the TLC plate, and an infrared radiating device.

Jupille and Perry (12) have reported that spot placement and size are independent of spotting technique in PMD. The spots produced by this method are uniformly and reproducibly compact regardless of the initial top-to-bottom spread, and identical spots arrive at the same level on the TLC plate regardless of initial displacement of the spot origin from the plate edge. PMD reconcentrates and aligns both high-R_f and low-R_f spots. For spots with low R_f values, reconcentration during solvent removal is more effective than reconcentration during mobile-phase advance.

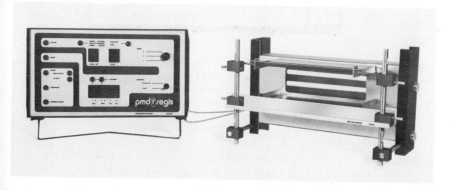

Figure 13.4. Instrumentation for programmed multiple development (PMD). Courtesy of Regis Chemical Co.

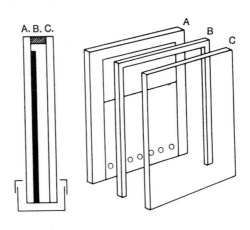

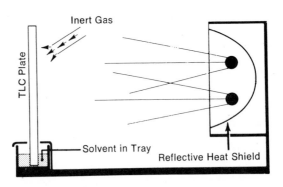

Figure 13.5.   Developer unit for PMD.   (a) TLC plate, (b) spacer, (c) cover plate.   The thin layer plate is about 5 in distant from and centered with respect to the radiator.   Courtesy of Regis Chemical Co.

The typical spot on a PMD chromatogram is an elliptical band with the top-to-bottom width smaller than the side-to-side width.
One important consideration must be kept in mind when using PMD.   Mobile phases with two or more components may change in composition during a long development and not provide reproducible separations.   This would be very likely to occur with mixtures of

solvents with widely differing boiling points, such as those containing diethyl ether. Using the attributes of PMD to their
fullest, it may be possible to effect separations with only single-component phases. Low-polarity solvents are generally recommended. Mobile phases used for conventional TLC cannot usually
be used for PMD because of such factors. A direct correlation
between the separations obtained by the two procedures should
therefore not be assumed; it will have to be proven experimentally.

It has also been generally found that the harder layers of
silica gel plates made with polymer binders are superior for PMD
development to the softer silica gel G plates with gypsum binder.

Jupille and Perry (13) have published a complete review of
PMD.

13.3 PYROLYSIS AND TLC

Rogers (14) needed a simple method for studying higher molecular
weight pyrolysis products as a continuous function of temperature.
He devised a method that involved programming a TLC plate across
the exit port of a pyrolysis cell as a function of the temperature
of the sample. The TLC plate was then developed in the usual manner, and reagent sprays specific for the compounds of interest
were employed for detection.

A pyrolysis cell was assembled from a 3/8-in stainless steel
Swagelok T-joint, using the 180° leg of the T for sample insertion. The 90° leg of the T is used for the carrier gas inlet and
must be positioned vertically. A 4.5-in section of 3/8-in stainless steel tubing was used as the pyrolysis chamber. This chamber was heated by a programmable tube furnace with an inside diameter of approximately 7/8-in. A 4-5 mm platinum or ceramic
sample boat may be used for weighing and pyrolysis, but a length
of platinum wire must be attached to the boat to enable sample
insertion and recovery of the boat. Control and readout thermo-
couples are welded to the 3/8-in tubing. The sample port may be
closed during a run with a brass (to prevent galling) Swagelok
plug. A 3/8-in to 1/4-in adapter can be used to fit a section of
1/4-in outside diameter stainless steel capillary to the cell to
increase the frontal velocity of the carrier gas downstream from
the sample. This section of tubing should be kept as short as
practical with the limitations imposed by the dimensions and heating capacity of the tube furnace. Rogers used a 2.5-in section of
0.1-in inside diameter. The sample boat should be as close to
this reduced diameter tubing as possible.

The TLC plate is vertically held within 1 mm of the end of
the stainless steel capillary and is drawn past the orifice on a
trolley as a function of the sample temperature. This movement
may be effected by extending the pen carriage of any suitable re-

corder. Rogers (14) found that a recorder with a variable span
that can be set to move exactly 5 in. for 20.65 mv was ideal. Such
an instrument, operating from a chromel-alumel readout thermo-
couple at the same position versus an ice reference, can be used
to push the plate 1 in. per 100°C.

Sample is applied along a band 1.5 in. from the top of the
plate, starting no less than 1 in. from the side of the plate
when the recorder is zeroed. The starting point is marked corresponding to recorder zero. The recorder is turned on, and the
furnace is heated at the desired programmed rate. The plate
should be pushed along the trolley track by the recorder at a rate
proportional to the heating rate, thereby making it possible to
correlate a position on the plate with the sample temperature. At
the end of the pyrolysis the plate is allowed to cool, inverted,
and developed in the chromatography chamber.

It is not necessary to use specific carrier gases for best
results. However, specific gases may be used to study particular
reactions and interactions during pyrolysis.

Desaga (Brinkmann, U.S. distributor) manufactures an apparatus for coupling high temperature vaporization with TLC. Called
a TAS (thermomicro, application, separation) oven or Tasomat, its
major use is for the separation of volatile substances from nonvolatile substances. The basic assembly is diagrammed in Figure
13.6.

The mixture to be separated is placed into a glass cartridge,
one end of which forms a capillary. This tube is inserted into an
oven that has been preheated to the desired temperature. The maximum is 350°C. Within 15-90 sec the escaping volatile substances
are transferred onto a TLC plate mounted on the oven and facing
the open end of the capillary. A sideways movement of the plate

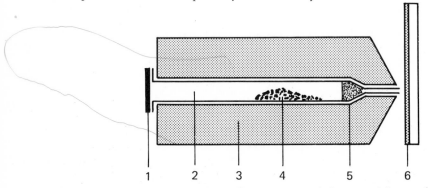

1	2	3	4	5	6

Figure 13.6. Cross-section of TAS oven with cartridge and TLC
plate. 1. Seal, silicone disk. 2. Glass cartridge. 3. Heating
block. 4. Sample. 5. Quartz wool. 6. TLC layer. Courtesy of
Brinkmann Instruments.

may result in additional prefractionation into low and high boiling constituents.

Stahl (15-17) has used the TAS procedure in the investigation of complex substances such as plant material, detergents, and shoe polish. He calls the entire procedure "thermofractography" (TFG), and the chromatogram after development a "thermofractogram." The chromatogram shows the substances separated by temperature on the abscissa, and separated by their chromatographic behavior on the ordinate. A hypothetical thermofractogram is shown in Figure 13.7. The instrument is shown in Figure 13.8.

Because of their high polarity, many high molecular weight natural and synthetic substances cannot be separated directly by chromatographic methods. By reproducibly degrading these substances to known lower molecular weight compounds, the original substance may be identified by a "structural unit analysis." The best procedures are thermolysis (up to 500°C) and pyrolysis (500-1200°C) (17).

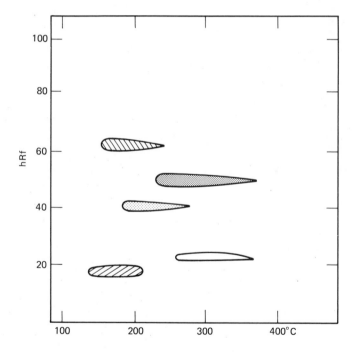

Figure 13.7. Thermofractogram. Substances separated by temperature are on the ascissa, and separated by their chromatographic behavior on the ordinate.

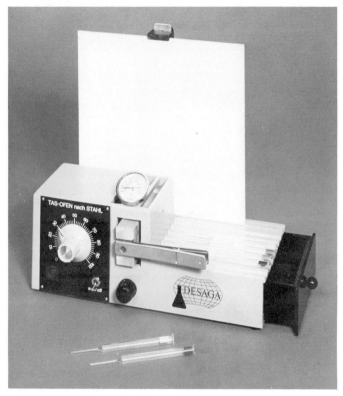

Figure 13.8.   TAS oven.   Courtesy of Brinkmann Instruments.

13.4 BIOAUTOGRAPHY

Bioautography is often the only way to detect antibiotics that
have been separated on paper or thin layer chromatograms. It is
based on the biological effects of the substances to be detected.
In general, these effects can be the inhibition or promotion of
growth of the organism that is exposed to the zone that was separated on a paper or thin layer chromatogram. Zones of inhibition
in paper chromatographs and zones of growth in thin layer chromatograms make possible the detection of antibiotics. Several
hundred antibiotics have been discovered in the past thirty years
Owing to the wide variations in chemical properties of these compounds, bioautography is the only general method for their detection on thin layer chromatograms.

The beginnings of bioautography appear in the work of Goodal

and Levi (18), who applied it to paper chromatograms. Fischer and
Lantner (19) and Nicolaus et al. (20) used it fifteen years later
for detecting antibiotics separated on thin layers.

In the first steps in the investigation of an unknown antibiotic, when it is not yet available in pure form, bioautography
becomes a method for classification of crude antibiotics by thin
layer chromatography. The method was developed for identifying
the antibiotics contained in a crude mixture during the early
stages of isolation of these compounds. The method attempted to
assess rapidly the probability that the antibiotic in question was
an already known one. The method was used as a screening technique, and 84 known antibiotics were investigated. The procedure
uses three solvent systems to yield four groups of antibiotics.
These are subsequently divided into 15 subgroups by use of 11 additional solvent systems.

For these studies, Eastman Chromagram sheets with silica gel
6060 were deactivated in air at room temperature (50-65% relative
humidity) for 24 hr before use. The chromatograms were developed
in paper-lined chambers in the usual way after sample application.
After development, the chromatograms were scanned under ultraviolet light and then subjected to bioautography.

The dried chromatogram was placed directly on a sheet of filter paper over a seeded agar gel slab. Incubation was then carried out at 37°C for 18 hr. Inhibition of the growth of specific
organisms then could be seen on the agar in the areas where the
antibiotic had separated.

There are a number of ways by which bioautography can be performed with TLC. Nicolaus et al. (20) prepared an agar medium inoculated with the test organisms to which triphenyltetrazolium
chloride had been added as a dehydrogenase inhibitor. This agar
medium was poured over the TLC and incubated in the usual manner.
The zones of separated antibiotic will, after a defined incubation
time, show up as areas of no growth.

Kline and Golab (22) first sprayed an agar solution onto the
TLC to solidify it. Then the inoculated agar, cooled to 48°, was
poured directly over the surface of the prepared plate. Bicket et
al. (23) simply pressed the TLC directly on the seeded agar slab.

For chromatograms that are friable and contain no binder, a
special technique was described by Meyers and Smith (24). A wet
sheet of Whatman No. 1 paper is centered over a glass plate the
same size as the TLC. The filter paper tabs that extend beyond
each end are folded back over the glass plate, and the assembly is
placed on top of the sorbent of the TLC. A sandwich results, with
the sorbent and filter paper between glass plates. The filter
paper tabs are folded back over the ends of the chromatographic
plate on the bottom. The latter is removed, leaving the filter
paper stuck to the glass support by means of the tabs folded over

it. The sandwich is then positioned on the surface of the prepared agar, which has been seeded with the specific organism.
With *Streptococcus lactis*, which grows irrespective of the presence of oxygen, results can be had overnight after incubation at
37°C, with the TLC and filter paper lying on the agar surface.
Antibiotics that are active against gram-negative organisms and
antibiotics active against fungi can also be detected when suitable test organisms are inoculated in the agar.

Meyers and Erickson (25) showed that gram-positive organisms
are noninhibitory for *Streptococcus lactis*, and the latter can be
replaced by *Staphylococcus aureus* when the agar is supplemented
with triphenyltetrazolium chloride and potassium nitrate. The
potassium nitrate is an oxygen donor in termal respiration;
Staphylococcus aureus will then grow on agar plates covered with
the TLC during incubation.

The methods described so far are based on inhibition or promotion of growth in the areas where the antibiotic has separated
on the TLC placed on the agar. Hamilton and Cook (26) described
the use of the phytopathogenic bacterium *Xanthomonas primi* for
studies in location of antibiotics separated on TLC. The organis
does not reduce tetrazolium dyes and is an obligate aerobe. It
hydrolyzes gelatin and starch and produces acid from several sugars. These characteristics can then be used as indicators of
growth when specific color reagents are used on the agar plates.

Zuidweg et al. (27) applied bioautography to layers of
Sephadex used as a sorbent for TLC. After development and drying
the chromatogram was pressed on the seeded agar plate covered wit
a sheet of lens paper. After contact for 30 min the TLC was removed and the lens paper taken off carefully. The agar plate was
then incubated at the conditions optimal for the test organism.
The technique of pouring seeded agar over the developed chromatogram can result in the antimicrobial substance spreading over the
plate to produce much larger areas of inhibition. This could be
overcome as mentioned by spraying a base of the agar over the sor
bent before pouring the main portion of the agar on the plate.
Pressing the plate on the agar requires the use of a microorganis
that can grow anerobically.

One way to overcome some of these disadvantages uses agar bu
depends on formation of a "lawn" on the surface of the agar. The
surface of the hardened but unseeded agar is inoculated with a
heavy suspension of the organisms. After incubation the confluen
growth areas are raised on the surface of the agar, and zones of
inhibition appear clear and depressed. If the colonial form is
opaque or pigmented, the contrast between zones of growth and inhibition is very noticeable against the background of the silica
gel. The spreading effect that results from pouring agar on the
sorbent can be overcome by laying preformed sheets of agar on the

surface of the sorbent.

Perlman et al. (28) have developed bioautography as a means of bioassay of cytotoxic substances. The differential agar-diffusion procedure for assessing antitumor compounds is based on the cytocidal effects of the test compound or zones separated on TLC. Eagle's KB cells, Earle's L cells, and L-1210 cells were suspended in Waymouth's Medium GM752/1 supplemented with calf serum for use as the assay medium. After development to separate the compounds of interest, the TLC was placed on the agar surface for 40 min and then removed. The plates were then incubated for 18 hr at 37°. For detection, the plates were flooded with vital stain and the dye reduction noted. The assay technique measures only cytocidal activity and is based on inhibiting cells from oxidizing glucose.

For preparative TLC, Lefemine and Hausmann (29) located colorless neutramycin on the chromatogram by making a "print" of the chromatogram with filter paper while the layer was still wet with the mobile phase. A sheet of glass was used to make even contact with the support. Enough of the separated antibiotic is transferred to the paper after contact for 10 min. The paper is then rid of the solvent and pressed against the agar plate. After the bioautography, the desired zones can be located on the preparative plate for further analysis or elution.

Homans and Fuchs (30) described a method for detection of fungistatic substances directly on the TLC. After development and drying of the chromatogram to remove solvent, a conidial suspension of a sensitive fungus in a suitable medium is sprayed directly on the layer. The thin layer plates are then incubated in a moist atmosphere for 2-3 days at 25°C. The positions of the zones of inhibition of growth are measured. From the quantitative standpoint, bioautography may be helpful, but the relatively high variability of biological quantitation must be considered. Consequently, quantitative usage has seen little practice.

Cephalosporin C and related compounds can be determined according to the method of Miller (31). The chromatogram was subjected to bioautography with _B. subtilis_ as the test organism. After incubation the maximum widths of the zones of inhibition were measured. The average maximum diameters for each of three standard samples were plotted against the amount of antibiotic on semilogarithmic paper. As the relative specific activities for the various compounds were known, it was possible to determine the amount of cephalosporin C, the deacetylated metabolite, and the lactone simultaneously in an unknown sample from the single standard curve for the parent compound. Reaction mixtures containing cephalosporin C and 7-aminocephalosporanic acid were analyzed for the acid after separation. The chromatogram was air dried and dipped in a solution of phenylacetyl chloride to con-

vert the reactive zones to the corresponding phenylacetate. This
chromatogram was then bioautographed using *Sarcina lutea* as the
test organism.

McDonald et al. (32) reported that the critical parameters
that must be controlled for quantitative bioautography include
the preparation of the TLC layer, sample application, mobile phase
selection, and plate storage. The microbiological conditions that
must be controlled include microorganism choice, organism storage,
preparation of medium, layer thickness, and dye systems for visualization. With careful control of these factors, bioautography
can be routinely used in the assay of lasalocid in diverse biological samples at concentrations of 0.5 ppm and higher.

Since it should be possible to control the factors of variability, it appears that bioautography may see more use in the
future. The availability of more uniform plates, media,prepack-
aging, and test organisms along with careful control of conditions
may make it possible to improve the methodology in this area of
investigation. This is important when the method is the only
available one, particularly in the search for new antibiotics.

13.5 ENZYME INHIBITION TECHNIQUES FOR USE WITH TLC

Thin layer chromatography has seen wide use in pesticide analysis.
Enzyme inhibition reactions have been applied to paper and thin
layer chromatography to detect some organophosphorus pesticides
(32-36) and carbamates (36). More recent work by Mendoza et al.
(37,38) have refined the methodology to the point where it is more
reproducible and has become generally accepted for detection of
pesticides from biological samples.

The method is based on the inhibition of the hydrolysis of
substrate esters by the pesticide. The pesticide combines with
the active sites on the enzyme to prevent its reaction with the
substrate ester, and no chromogenic product will be formed. On
the TLC plate, the areas where the pesticides are located will
appear as colorless spots against an intensely colored background.
The enzyme solution is sprayed on the TLC plate after development.
After a short time, the substrate is sprayed on the plate. If
the pH indicator is used, the pesticide zones appear as colored
spots against a colorless or lightly colored background.

Geike (39) used the inhibition of trypsin for the determination of chlorinated pesticides. Later (40) he used phosphatase
and naphthylphosphate as a substrate. Other examples of substrate
are 1-naphthyl acetate, which is detected by reaction with a
diazonium salt after hydrolysis. Indoxyl acetate gives a fluorescent reaction product if it is hydrolyzed.

The detection limits for each pesticide or group of pesti-

cides vary with the type or source of the enzymes. This was
shown by Winterlin et al. (41) and by Mendoza and co-workers (37,
38, 42). The method is sensitive to as little as 0.5 ng of pesticide (carbaryl-using pig liver enzyme). Beef liver esterase was
much more sensitive to organophosphorus pesticides than pig liver
esterase. The opposite was true for carbamates. A marked difference in specificity was noted for inhibition of beef and pig
esterases due to malathion and parathion (42). In contrast, carbaryl was detected at 0.1 ng by the frozen extract of pig liver
and at 0.5 ng by the freeze-dried extract of pig livers or frozen
extract of steer livers. Researchers using the methodology should
check these enzyme preparations.

In other work by Mendoza et al. (38), the specificity of the
enzyme interactions was best demonstrated by the carbaryl-pig
liver esterase, demeton-pig or monkey esterases, and by the
demeton compounds-chicken liver esterase combinations. Carbaryl
at the 1 ng level and demeton at the 50 ng level inhibited the
pig liver esterases regardless of the method of enzyme extraction.
The esterases from the other species were inhibited only by carbaryl exposed to bromine. Demeton was inhibitory to monkey esterases extracted with water or Tris buffer.

The effect of dilution of the enzyme on the sensitivity of
detection of some carbamates and organophosphorus pesticides is
pronounced. Diluted enzyme sprays gave more sensitive detection
than did concentrated sprays. The difference can be as much as
five fold. This is due to the fact that in a concentrated enzyme
solution the number of active sites of the enzyme is greater than
the pesticide could block. The effect of pH of the sprays must
also be controlled. No organophosphorus pesticides were detected
at pH 5.3 with beef liver esterases and 5-bromoindoxy acetate. A
pH of 8.3-9.1 appeared maximal (37).

The method has been used in a number of forensic applications. Heyndrickx et al. (43), reviewed cases of parathion poi-
soning in humans. They determined the distributions of parathion in various organs using fresh horse plasma and 1-naphthyl
acetate after separating the pesticide by TLC. The detection
limits varied from 0.05 µg for paraoxon to 25 µg for fenthion.

Some general details of the method as outlined by Mendoza
and Shields (42,44) are given in the following sections; they must
be modified according to the material from which the pesticide is
to be extracted.

13.5.1 Extraction of Plant Samples

The sample (50 g) and 50 g of anhydrous sodium sulfate were blend-
ed with 150 ml of chloroform:methanol (90:10) for five min in a

Waring blender. This was filtered through Whatman No. 1 paper.
The solids were rinsed twice with the methanol-chloroform solution. Sodium sulfate was added to remove water. Then the fil-
trate was diluted to 200 ml in a volumetic flask.

Appropriate aliquots were concentrated to obtain equivalent
amounts of plant extract. Equivalent amounts of selected pesticide standards were spotted over the plant extracts on TLC plates
The resulting chromatograms were then compared to determine the
effects of one extract on the enzyme.

13.5.2 Preparation of Enzymes

In a Virtis homogenizer, blend 50 g of fresh livers for 2 min
with 180 ml of cold distilled water or Tris buffer solution (pH
8.3) containing nicotinamide, each at 0.01 M. Centrifuge the
homogenate at 2000 G at 4°C for 5 min. Freeze the supernatant in
13×100 mm test tubes. The frozen extracts can be used up to
several months after preparation.

To prepare the spray, dilute 1 part of the thawed extract
with 8 parts of 0.05 m Tris buffer solution before use. The dilution may be varied depending on the intensity of the indicator
used on the TLC plate. For the preparation as described, one
tube diluted 8 times is enough to spray 6 plates.

13.5.3 TLC Procedure

The plates used as recommended by Mendoza et al. (44) for most of
their work were silica gel G-HR coated to a thickness of 750 μ.
Silica gel H and G and aluminum oxide DS-5 have also been used.
The particular pesticide to be used will determine which one of
the sorbents may be used. The best one for a particular purpose
must be found by trial. The plates were heated 1 hr at 110° for
activation before use.

The pesticides and residues were applied on the plates (10
μl). The solutions were made up in alcohol or acetone to contain
from 1 to 100 ng depending on the inhibition expected. Carbaryl
is sensitive at the 1 ng level, while dimethoate requires 10,000
ng for inhibition. The amount applied will depend on the pesticide under consideration.

Most pesticides will be resolved after development with a
solution of acetone in hexane (20:80). A mobile phase travel of
15 cm is sufficient. When the plates are dry (20-30 min), the
substrate solution is evenly sprayed with the enzyme solution until the plate is just thoroughly wet. Allow the plate to dry before spraying with the chromogenic solution. Indophenyl acetate
(1 mg/ml in acetone) or 5-bromoindoxyl and indoxyl acetate dissolved in ethanol (0.7 mg/ml) and in acetone (5 mg/ml), respect-

ively, can be used as substrates for organophosphorus and carbamate pesticides. If the pesticide has combined with the active sites on the enzyme, clear colorless zones will appear on the dark background of the plate.

Mendoza et al. (45) also investigated the effect of bromine and UV light on the pesticides. Differences in ability to inhibit esterases and in migration rates on TLC were found with the pesticides and their various metabolites. Some of the pesticides showed greater inhibition on exposure to bromine on the TLC plate prior to development. These differences in inhibition among the various pesticides could be used in identification procedures based on the TLC-enzyme inhibition technique.

These TLC methods in combination with enzymic techniques are versatile and sensitive for detecting pesticides. The combined technique is useful in the analysis of pesticide residues in foods and the environment, in metabolic and forensic investigations, and in analysis of samples with interferences that are too great for other methods. Up-to-date reviews of the present status of this technique have been compiled by Mendoza (46,47).

REFERENCES

1. R. A. deZeeuw, Anal. Chem., *40*, 2134 (1968).
2. R. A. deZeeuw, J. Chromatogr., *48*, 27 (1970).
3. R. A. deZeeuw, J. Pharm. Pharmacol., *20*, 54S (1968).
4. F. Geiss and H. Schlitt, Chromatographia, *1*, 392 (1968).
5. S. Sandroni and H. Schlitt, J. Chromatogr., *52*, 169 (1970).
6. A. V. DeThomas and F. Pascual, J. Chem. Ed., *46*, 319 (1969).
7. V. V. Rachinskii, J. Chromatogr., *33*, 234 (1968).
8. S. Turina, Z. Soljic, and V. Marjanovic, J. Chromatogr., *39*, 81 (1969).
9. S. W. Mayer and E. R. Tompkins, J. Amer. Chem. Soc., *69*, 2866 (1947).
10. S. Turina and V. Jamnicki, Anal. Chem., *44*, 1892 (1972).
11. J. A. Perry, K. W. Haag, and L. J. Glunz, J. Chromatogr. Sci., *11*, 447 (1973).
12. T. H. Jupille and J. A. Perry, J. Chromatogr. Sci., *13*, 163 (1975).
13. T. H. Jupille and J. A. Perry, Science, *194*, 288 (1976).
14. R. N. Rogers, Anal. Chem., *39*, 730 (1967).
15. E. Stahl, J. Chromatogr., *37*, 99 (1968).
16. E. Stahl, Analyst, *94*, 723 (1969).
17. E. Stahl and T. Herting, Chromatographia, *7*, 637 (1974).
18. R. R. Goodall and A. A. Levi, Nature, *158*, 675 (1946).
19. R. Fischer and H. Lantner, Arch. Pharm., *294*, 1 (1961).

20. R. J. R. Nicolaus, C. Coronelli, and A. Binaghi, Farmaco (Pavia) Ed. Pract., *16*, 349 (1961).

21. A. Aszalos, S. Davis, and D. Frost, J. Chromatogr., *37*, 487 (1968).

22. R. M. Kline and T. Golab, J. Chromatogr., *18*, 409 (1965).

23. H. Bickel, E. Gaumann, R. Hutter, W. Sackman, E. Vischer, W. Vosen, A. Wettstain, and H. Zahner, Helv. Chim. Acta, *45*, 1396 (1962).

24. E. Meyers and D. A. Smith, J. Chromatogr., *24*, 129 (1967).

25. E. Meyers and R. C. Erickson, J. Chromatogr., *26*, 531 (1967).

26. P. B. Hamilton and C. E. Cook, J. Chromatogr., *35*, 295 (1968)

27. M. H. Zuidweg, J. G. Oostendorp, and C. J. K. Bos., J. Chromatogr., *42*, 552 (1969).

28. D. Perlman, W. L. Lummis, and H. J. Griersbach, J. Pharm. Sci., *58*, 633 (1969).

29. D. V. Lefemine and W. K. Hausmann, Antimicrob. Ag. Chemother., 134 (1963).

30. A. L. Homans and A. Fuchs, J. Chromatogr., *51*, 327 (1970).

31. R. P. Miller, Antibiotic Chemother., *13*, 689 (1962).

32. A. McDonald, G. Chen, P. D. Duke, A. Popick, and R. A. Saperstein, Abstracts, 169th National Meeting, American Chemical Society, Philadelphia, Pa., April 7, 1975.

33. W. P. McKinley and S. L. Read, J. Ass. Off. Agric. Chem., *45*, 467 (1962).

34. S. P. McKinley and P. S. Johal, J. Ass. Off. Agric. Chem., *46*, 840 (1963).

35. R. Ortloff and P. Franz, Z. Chemie, Lpg., *5*, 388 (1965).

36. J. J. Men and J. B. Bain, Nature, *209*, 1351 (1966).

37. C. E. Mendoza, P. J. Wales, H. A. McLeod, and W. P. McKinley, Analyst, *93*, 34 (1968).

38. C. E. Mendoza, D. L. Grant, B. Braceland, and K. A. McCully, Analyst, *94*, 805 (1969).

39. F. Geike, J. Chromatogr., *52*, 447 (1970).

40. F. Geike, J. Chromatogr., *61*, 279 (1971).

41. W. Winterlin, G. Walker, and H. Frank, J. Agr. Food Chem., *16*, 808 (1966).

42. C. E. Mendoza and J. B. Shields, J. Chromatogr., *50*, 92 (1970).

43. A. Heyndrickx, A. Vercruysse, and M. Noe, J. Pharm. Belg., 127 (1967).

44. C. E. Mendoza and J. B. Shields, J. Ass. Off. Anal. Chem., *54*, 507 (1971).

45. C. E. Mendoza, P. J. Wales, D. L. Grant, and K. A. McCully, J. Agr. Good Chem., *17*, 1196 (1969).

46. C. E. Mendoza, J. Chromatogr., *78*, 29 (1973).

47. C. E. Mendoza, Residue Rev., *50*, 43 (1974).

CHAPTER 14

The Combination of
Thin Layer Chromatography
with other
Analytical Techniques

14.1   INTRODUCTION

Thin layer chromatography is a separation process that is useful
in a number of ways.   The R_f values obtained for separated substances and the comparison of these substances with known compounds separated on the same plate serve as useful guides for the
identification of an unknown but are not sufficient for absolute
identification of the unknown.

The careful combination of TLC with other chromatographic
techniques and with appropriate identification techniques will provide sufficient information about a substance to lead to its identification.   Column, high pressure liquid, and gas chromatography
are the three most widely used chromatographic techniques in conjunction with TLC, with paper chromatography and electrophoresis
is being used less frequently.   Chromatographic procedures are
excellent separation methods for complex mixtures but are of
little value by themselves in the identification of the structure
of the separated substance.   Hand in hand with the chromatographic
procedures are the identification methods, which are useful for
elucidating the structure of a substance, but these methods require relatively pure material, which the chromatographic proce-

347

dures can provide.

Identification techniques used include:

1. Chemical reaction such as colored product formation, derivatization, chemical pyrolysis. See Chapter 7 for such techniques.

2. Spectrographic procedures such as IR, mass spectroscopy, and NMR.

3. Physical methods such as melting point, sublimation, polarography.

4. Biological or physiological methods such as testing an antibiotic, pesticidal, or bioautography effect. See Chapter 13.

The chromatographic and spectrographic methods will be discussed in depth, as they are generally the most widely used.

14.2   COLUMN CHROMATOGRAPHY

Some of the correlations and uses of the combination of column chromatography (CC) and TLC have been reviewed by Janak (1,2) and by Schlitt and Geiss (3). Since both are forms of liquid chromatography (that is, they employ a liquid mobile phase) separation mechanisms for TLC and column chromatography become the same if the mobile phase and sorbent are the same. Often silica gel or alumina is used as the sorbent in column chromatography, for instance.

Because TLC has many merits and advantages, it can be used as a pilot technique to select the best combination of mobile phase and sorbent for a column separation. Boshoff et al. (4) have recently coupled TLC with high performance liquid chromatography (HPLC). Solutes from the column were transferred by means of a fine steel capillary tube (0.25 mm I.D.) onto 5×20 cm chromatoplates that were moving under the column at a speed of 2 cm/min. The plate was being heated, and a small vacuum line was placed above the solute zone at the same time, to evaporate the elution solvent. With steroids, Boshoff et al. were able to detect 60 ng/spot, and with the chlorinated pesticide methoxychlor, 0.6 ng/spot. A spectrofluorimeter was used for detection, after the production of fluorescent derivatives *in situ*. Snyder (5) claims that the resolving power of columns in modern, rapid CC is greater than that of a TLC plate, which is limited by its maximum length. The relative merits of TLC compared to GC are:

1. Most separated substances may be detected visually, often with a specific color reaction. Many CC detectors are not specific or sensitive enough for many applications.

2. The amount of time required to change a set of separation conditions is very short in TLC compared to CC. It usually
only takes about 0.5 hr to change conditions in TLC, compared to
several hours in CC. Column reequilibration after changing the
composition of the mobile phase takes at least 1 hr.
    3. With TLC it is possible to do many separations in 20 min.
    4. The major advantage of TLC is its great economy in ma-
terials and time, which can be used to advantage in itself or to
determine the optimum set of conditions for a column separation.

After the separation conditions have been determined and the
column is being developed, the composition of the eluent may be
monitored by applying aliquots taken at given time or volume intervals to a TLC plate and developing and visualizing it. Results
will show which column fractions may be combined, discarded, or
otherwise acted upon.
    The reverse procedures may also be used; that is, an eluted
zone from a TLC plate may be placed in and developed through a
column.

14.3 GAS CHROMATOGRAPHY

There are two major ways in which thin layer chromatography may be
combined with gas chromatography: indirectly and directly. The
majority of TLC/GC work is done indirectly: The sample is developed on a TLC plate, the separated areas may be tentatively
identified, and desired zones are eluted from the plate, concentrated, derivatized, or otherwise modified, and then subjected to
separation and identification by gas chromatography. A number of
applications employing this general procedure may be found in the
literature; these include analysis of steroids (6-14), cardiac
glycosides (digoxin) (15), amines (16) nitro compounds (17,18),
imidazoles (19), organophosphorus compounds (20), fat-soluble
vitamins (21), water-soluble vitamins (22), insecticides (23),
pesticides (24,25), *Cannabis* hallucinogens (26), drugs of abuse
(27,28), antioxidants (29), and unsaturated fatty acid methyl
esters (30-33).
    Two examples are chosen to illustrate the indirect combina-
tion of TLC with GC. Working with urinary estrogens, Touchstone
et al. (7) extracted the steroid conjugates from an entire 24 hr
urine sample and hydrolyzed the extract to release the steroids.
The hydrolysate was extracted several times with ether; then the
ether extract was washed and subjected to an extraction to obtain
the phenolic steroids (estrogens) present. The estrogen extract
was applied to the origin of a silica gel G plate, and known
estrogen standards were applied next to the extract. The plate

was developed in 15% acetone: isopropyl ether and dried. The
sample lanes were masked with a clean glass plate, and the reference standards were visualized with ferric chloride-potassium
ferricyanide. The sample zones corresponding to the standard
zones were individually scraped off into micro separatory funnels
containing 15 ml of pH 4 water and extracted three times with 15
ml of ether each time. The ether extracts were washed with 5 ml
of 8% NaHCO$_3$ and twice with 5 ml of water. The ether was evaporated and the residue transferred to a conical tipped tube. A
known amount of suitable solvent such as methanol or tertiary
butanol was added to the residue, and a suitable aliquot was used
for quantitative gas chromatography.

The TLC separation serves to isolate sample components of
interest from the unwanted sample contaminants and provide a
tentative identification for them. The gas chromatography provides further separation, identification, and, in this case, quantitation of the desired estrogens. The two methods complement
each other well for optimum results.

As a second example of the utility of the combination, consider the work of Privett et al. (30-33). Working with methyl
esters of unsaturated fatty acids, they separated them on silica
gel layers impregnated with silver nitrate, which provides a separation according to the degree of unsaturation. The zones were
eluted and reduced by ozonolysis, and the resultant compounds were
analyzed by gas chromatography to obtain additional information
for identification.

The second major way in which TLC is combined with GC is by
direct physical coupling of the GC instrument with the TLC plate.
This method has been reviewed by Kaiser (34) and by Janak (35,36),
one of the pioneers in this area (37-39). Kaiser points out three
advantages to this coupling:

1. It permits a double chromatographic separation in two
dimensions. Results are obtained by comparison of the gas chromatogram with the GC/TLC chromatogram.

2. The combination offers the possibility of individual and
multiple qualitative identification as well as quantitative and
qualitative determination of individual compounds.

3. The method is a most critical control procedure. Of major value is the disclosure of contradictions in the qualitative
results comparing the gas chromatogram with the GC/TLC chromatogram. Such results will allow any quantitative results to also
be checked. Overall the two systems complement each other, because the separations are dependent on different characteristics
of the sample components.

The first workers to couple a TLC plate with a gas chromatography were Casu and Cavallotti (41). They positioned the plate under the effluent of the chromatograph, driving it past by means of a gear train from a potentiometric strip chart recorder. The plate retained the sample components, and *in situ* chemical derivatization was used for identification. The separation capabilities of the TLC process were not actually utilized, as the plate served only as a trap.

Janak was the first to use the TLC separation process after coupling a plate to the gas chromatograph (37-40). Tumlinson et al. (42,43) manually moved the plate past the GC effluent to collect the sample components, and formed derivatives *in situ* before developing the plate. Curtius and Muller used independent plate movement and the coupling procedure for steroid analysis (44). Humphrey has examined volatile oils by this procedure (45).

The only commercial instrument on the market at present that is designed to move a TLC plate past the effluent of a gas chromatograph is Camag's Diochrom, illustrated in Figure 14.1. Ten continuous plate speeds are possible: 1, 2, 5, 10, 20, 50, 100, 200, 500, and 1000 mm/min, or 200, 100, 40, 20, 10, and 4 min (120, 60, 24, and 12 sec) per 20 cm plate. Stepwise advancements of 15, 20, and 25 mm can also be selected.

The plate movement is initiated by the operator when the first peak of interest is observed on the recorder used with the gas chromatograph. The sample components can be automatically collected thereafter on either side of the plate. In the stepwise mode, the plate movement may be monitored by the deflection of the recorder pen. Available optionally with the unit is a heated connecting line with a temperature control for high-boiling sample components.

It is desirable to move the plate continuously past the effluent and subsequently to be able to compare the visualized areas on the TLC plate with the peaks produced by the chromatograph recorder. Casu and Cavallotti (41) used the drive gears for the paper on the recorder to turn other gears, which moved the plate. The effluent from the gas chromatograph was placed about 1 mm above the surface of the plate layer. Humphrey (45) used the recorder chart paper itself as the driving mechanism for the plate. The recorder was equipped with a flat shelf extending 40 cm at an angle of 45° from the point of emergence of the chart paper. The chart was allowed to run over the sloping shelf, and the TLC plate was fixed to the paper with a piece of adhesive tape. The open capillary end of the heated effluent stream splitter was positioned 2 mm above the plate surface. If a destructive detector such as the hydrogen flame detector is being used in the gas chromatograph, it is necessary to split the effluent emerging from the column before it arrives at the detector. A small portion, in this case 2%, is fed to the detector, the other 98% being al-

lowed to pass out to the TLC plate. This is the same principle
used during fraction collection in preparative gas chromatography
The effluent capillary to the plate was kept at a temperature of
150° for most applications, but 180° or 200° was necessary for
oils containing phenols or sesquiterpene alcohols.

Ruseva-Atanasova and Janak (40), working with fatty acid
methyl esters, logarithmically programmed the speed of the plate;
this resulted in linear and additive spacing of neighboring mem-
bers of a homologous series. Their work illustrates the utility
of the TLC/GC coupling. Gas chromatography of methyl esters on a
nonpolar stationary phase (the first-dimension separation) resulted in a nice separation of the fatty acids according to the number of carbon atoms in the molecule, regardless of the number of
double bonds. A second-dimension separation of the esters on a
silver nitrate-impregnated silica gel plate resulted in a separation according to their degree of unsaturation, regardless of the
number of carbon atoms. The logarithmic variations of the speed
of the plate past the effluent of the gas chromatograph ensured
that the identification of the individual compounds on the developed chromatogram became a matter of linear geometric orienta-
tion. This method simplifies the comparison of various lipids,
and of differences in their composition depending on their biological source.

Figure 14.1. Camag Diochrom instrument for the coupling of a
thin layer chromatography plate to a gas chromatograph. Courtesy
of Camag, Inc.

Ruseva-Atanasova and Janak point out that both the amount of silver nitrate in the silica gel and the polar component of the mobile phase influence the R_f values obtained. The amount of silver nitrate is important for optimum resolution of areas and for well-defined separations without tailing. Their best separations resulted from a minimum content of 1 g AgNO$_3$ per gram of silica gel. An increase in the amount of the polar component in the mobile phase accelerated the migration of the unsaturated acids up to a limit. Thereafter, tailing of the zones resulted. The optimum mobile phase was 70:30 petroleum ether-diethyl ether.

Figure 14.2 shows the chromatograms typically obtained from a coupling of TLC with GC. After the effluent is collected on the plate, developed, and visualized, it is placed alongside the GC recorder tracing (the gas chromatogram) such that the visualized zones correspond to the peaks. Starting points should be carefully marked before separation. Examination of the spots on the plate will indicate whether the GC peaks truly represent single components or whether inadequate separation is occurring in either dimension.

When the GC column is having its temperature raised by temperature programming, it is advantageous to move the TLC plate past the column effluent at a linear rate. Dilution of the deposited spot is then avoided.

A number of advantages are afforded by coupled or combination TLC/GC analysis that are not toally present with TLC or GC alone. The flexibility of choice of the column packing for the gas chromatography together with the limited choice of the layer material for the TLC plate is a major advantage. There are many GC column packings available for specific separations and an abundance of literature detailing extractions, purification procedures, and chromatographic conditions for a wide variety of compounds in many fields. The worker desiring to couple TLC with GC or to use them in combination has a wealth of information available for establishing a qualitative and/or quantitative procedure for a single substance or an entire class.

Many commercial suppliers of TLC plates, equipment, reagents, and reference compounds are also suppliers of columns, packings, liquid phases, reagents, equipment, and reference compounds for GC. Such suppliers include Analabs, Applied Science, and Supelco Co.

Another nice advantage to coupled analysis is the flexibility of detection afforded by the two chromatographic methods. General detectors such as the hydrogen flame and argon ionization and specific detectors such as the electron capture detector are available for GC detection according to the type of compounds being separated. This provides some flexibility for the identification of substances in the first-dimension (GC) separation.

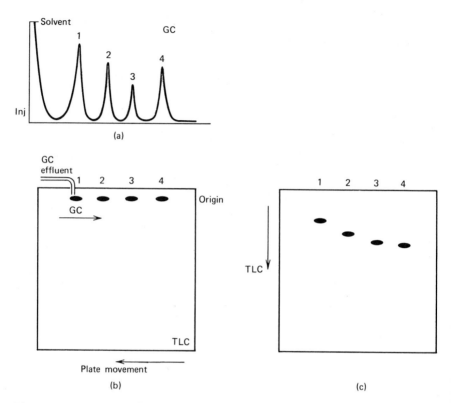

Figure 14.2.   Coupling TLC with GC.   (a) Gas chromatogram.   (b)
Spotted plate.   (c) Developed TLC chromatogram.

     After the separated substances are collected on the TLC
plate and separated in the second dimension, a variety of methods
may be applied for their detection and identification. Many visualization reagents are very specific in their chemical reaction
and provide a high degree of reliability in their use.   In addition to the qualitative identification afforded by the second-
dimension separation and detection procedures, *in situ* quantitative determination of a wide variety of compound classes is also
possible (46).   This is an additional benefit made possible by
sophisticated commercial spectrodensitometers that often saves
time, avoids losses, and affords greater sensitivity than elution
methods.
     Detectors commonly used in gas chromatography may also be

used for detection in TLC. This is not the same as coupled GC/
TLC, as no separation occurs in a second dimension; only the detector for a gas chromatograph is utilized. Cotgreave and Lynes
(47) detected separated substances on a TLC plate by vaporizing
them with a furnace and passing them to a suitable detector by
means of a gas stream. Padley (48) separated lipids on a thin
rod of high-temperature silica glass coated with silica gel G.
The rod was then passed through the flame of a flame ionization
detector, which combusted and detected the separation lipids.
The detector signal was amplified and recorded on a strip chart
recorder. Szakasits et al. (49) performed separations of a variety of compounds on 3 mm wide metal strips layered with silica
gel. After development, the strip was passed directly between
the nozzles of a dual-jet flame ionization detector. The resultant signal was passed to an electrometer, a recorder, and a digital integrator. The strip was passed through at a sufficiently
high temperature (300-450°) that all sample components were removed on the first pass. Complete experimental details on the
custom fabricated apparatus are given in the paper.

14.4 MASS SPECTROMETRY (MS)

The combination of TLC with a mass spectrometer or with a gas
chromatograph coupled to a mass spectrometer (GC/MS) is very useful for the unambiguous identification of separated substances.
The major factor limiting the use of such vital combinations for
identification has been the high cost of the GC/MS or MS systems.
A variety of instrumentation is now available over several price
ranges, and this will aid many workers in choosing a system to
fit their needs and budget.
     Lisboa (50) has shown how the combination of TLC with GC/MS
has proved to be indispensable for the positive identification of
steroids. Such a combination is valuable (1) for steroids with
similar polarities by both TLC and GC, but with different MS
fragmentation; and (2) for steroids with similar patterns of mass
fragmentation but different chromatographic mobilities. Many
examples of the usefulness of this technique in steroid identification are presented.
     Majer et al. (51) have combined TLC with high-resolution MS
in a highly sensitive rapid method of analysis for polycyclic
aromatic hydrocarbons. Such methods are of major importance in
the determination of air pollutants, since many of these particular compounds are carcinogenic and it is therefore necessary to
be able to determine very low levels. Majer et al. point out
that the sensitivity of the method is several orders of magnitude
lower than that of the fluorescence methods that are primarily

used for this type of compound.

Details of the quantitative analytical procedure are presented here as a guide for practical application of TLC with MS.

The mass spectra of the individual reference compounds of interest are recorded by evaporating a microgram quantity into the source of the mass spectrometer. On examination of the resulting spectrum, a characteristic ion is selected for the quantitative examination. With the polycyclic compounds this is usually the molecular ion, as it carries a high percentage of the total ion current.

In order to establish calibration curves, known quantities of the compounds of interest are chromatographed in the weight range required. For quantities greater than 10 ng of the polycyclic compounds, the position of the resulting spot may readily be determined by examination under UV light. For smaller quantities, or for compounds not visible under UV, the small quantity must be developed alongside a larger quantity of the same substance that is visualized. The location of the smaller quantity is then found with a straight edge. The spot is marked, scraped off the plate with a spatula, and transferred into a capillary centrifuge tube with the aid of a platinum scoop and a camel hair brush. The sample is then extracted from the layer material by stirring with 20-50 µl of solvent and centrifuging the resulting suspension.

A measured volume of this solution is transferred into the sample probe of the spectrometer. This is introduced into the vacuum system through the insertion lock and held in the cool part of the system until the solvent has evaporated. A reference compound, in this case heptacosafluoro-tri-n-butylamine, as introduced into the source by an alternative inlet for calibration of the mass scale of the instrument. The sample probe is then lowered into the heated part of the source to allow evaporation of the sample while the rise and fall of the ion current is being recorded. The area under this curve is directly proportional to the amount of substance evaporated into the ion source. A calibration curve may be plotted for each known compound.

After an unknown has been separated by TLC, it is eluted and carried through the same procedure. The quantity of the unknown may be obtained by measuring the area under its ion peak and relating this to the calibration curve obtained with the reference compound.

Majer et al. (51) obtained calibration curves that were linear between 10^{-6} and 10^{-12} g. Using the instrument gain and further amplification of the ion current, it is possible to lower the limits of detection still further. They note that the sensitivity may be further increased by decreasing the thickness of the sorbent layer on the TLC plate. A smaller volume of solvent would then be required for extraction.

It is generally necessary to elute the substance to be characterized out of the TLC sorbent before placing it into the spectrometer. If this material were placed directly into the instrument, it would soon become contaminated with sorbent dust, resulting in sporadic operation and eventual shutdown of the instrument
for cleaning.

Fetizon (52) isolated several milligrams of substance by
extraction from a large number of 250 μ thick plates and used
this for mass spectrometry.

Schwartz et al. (53,54), working with metabolites of the
drugs diazepam and chlordiazepam, subjected extracts from TLC
plates to mass spectrometry. Their spectra had strong background
noise, making it difficult to distinguish the signal from the compound being sought. Microfiltration through a membrane filter in
a Swinney adapter (see Chapter 12) might have been of value in
ridding the extracts of layer sorbent particles.

Boulton and Mayer (55) used TLC/MS for the quantitative
analysis of p-tyramines in biological extracts.

Szekely (56) has described a simple technique for preparing
an eluted substance for MS without the sorbent layer contamination
that leads to high background noise. Shown in Figure 14.3, it
consists of a glass capillary 80 mm long and 1 mm inside diameter,
such as an open-ended melting point capillary, which is packed
with finely powdered potassium bromide sealed in by glass wool.
The eluted sample in a small volume (50-100 μl) of volatile solvent is placed into a microcentrifuge tube, reaction vial, or
other such vessel. The capillary filled with the potassium bromide is placed into this vessel, and the sample solution is allowed to migrate up. An additional 30-50 μl of solvent is then added to the vessel and allowed to migrate up. This process serves
to concentrate the sample substance in the upper portion of the
potassium bromide, while the lower portion serves to filter out
the sorbent. Most of the lower part of the tube is broken off
and discarded. The upper portion of the capillary is introduced
through the direct inlet system of the instrument after the sample head has been suitably modified. This procedure requires at
least 1 μg of sample.

A microcolumn to hold the TLC sorbent to be eluted may be
made from a 230 mm long disposable Pasteur pipet (57), as in Figure 14.4. The sorbent is supported by a small plug of solvent-
washed glass wool, and this is eluted with 50-200 μl of solvent.
When elution is believed to be complete, the lower portion of the
column is broken off and the contained eluate is evaporated on
the tip of the solid sample probe of the mass spectrometer. Mass
spectra free of impurity peaks have been obtained with as little
as 5 μg of sample.

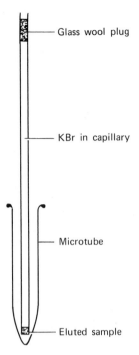

Figure 14.3. Microcolumn transfer technique (56). See text for explanation.

   Throughout any of these elution and preparation procedures, whether for GC, MS, IR, Preparative TLC, or other, it is imperative to employ scrupulously clean glassware and apparatus, the finest quality reagents and solvents, and an atmosphere and technique suitable for such detailed micro work. It is desirable, when it is known that isolation and elution are to be carried out, to prewash (predevelop) the TLC plate in a suitable solvent such as methanol. The solvent will carry impurities present in the layer to the top of the plate. If desired, the top 2-3 cm of the layer containing these impurities may then be scraped off and discarded before the plate is used. This procedure may be used for preparative plates also.
   Amos (58) has demonstrated the importance of using careful technique and high quality solvents, apparatus, and layer materials. He demonstrated the presence of a readily extractable plasticizer (acetyl tributyl citrate) in the plastic nozzle of a commercial vacuum cleaner type of layer-collection apparatus as described in Chapter 12. The plasticizer was extracted by con-

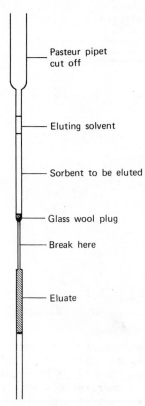

Pasteur pipet
cut off

Eluting solvent

Sorbent to be eluted

Glass wool plug

Break here

Eluate

Figure 14.4.   Elution column made from long Pasteur pipet (57).

tacting the dry sorbent being sucked into the extraction thimble.
Removing the plastic nozzle solved the problem, but the glass
plates were scratched.   The researcher thought this undesirable
and subsequently used a spatula to scrape off desired zones.
    The solvent-extractable impurities in commercial TLC grade
silica gels were measured by Amos.   The results are presented in
Table 14.1.   These results indicate that Merck silica gel HR contains the least amount of extractable material and that the H
series of silica gel contain less impurities than silica gel G or
the other commercial silica gels.   The researcher eluting substances from silica gel layers should work with the H and HR series gels whenever possible.   Amos notes that the amounts of
ethanol-soluble substances are higher than the acetone-soluble
materials because of the increased solubility of inorganic ad-

TABLE 14.1   SOLVENT-EXTRACTABLE MATERIAL IN TLC GRADE
SILICA GELS (58) (mg/100 g)

Source	Acetone Solubles	Ethanol Solubles
Merck silica gel		
G	19.2	42.2
H	3.6	28.4
HF 254	4.4	34.0
HF 254 + 366	4.6	31.8
HF 254 silanized	15.4	88.8
PH 254 silanized	7.6	63.0
HR pure	1.0	28.4
F 254	38.8	92.2
Whatman Chromedia	11.4	23.4
Woelm silica gel	38.2	374.0
B.D.H. silica gel	10.6	73.8
M&B Chromalay	23.2	37.6
Blank on solvent and filter	>0.1	>0.1

ditives (binders, phosphors, etc.) and silica. Accordingly, polar solvents such as alcohols are clearly unsuitable for the extraction of TLC spots for IR or MS examination.

The purity of commercial solvents was also examined by Amos.
He points out that to obtain a spectrum of 10 μg of material
eluted from a TLC spot, no more than 0.5 μg of total impurity can
be tolerated. This equates to not more than 0.5 ml of solvent
containing 1 mg/liter of nonvolatile residue being used to elute
a spot (and this does not consider what is eluted as impurity
from the layer material itself). Volumes of a number of commercial solvents were evaporated and the residues weighed to determine if such criteria could be met. IR spectra were also made of
the residues.

It was found that all batches of analytical reagent grade
chloroform were contaminated by phthalate esters. This may be
due to contact with plastic tubing during the bottling process.
Ether residues contained the inhibitor n-propylgallate. Ether
is not recommended for elution; acetone is a suitable substitute.
Expensive high purity solvents bottled with plastic or Teflon-
lined caps were found to be generally suitable. When these are
not available, it is suggested that analytical reagent grade solvents be glass distilled into glass-stoppered bottles of high

quality glass.

14.5   INFRARED SPECTROSCOPY

Using the previously described elution techniques and taking suitable precautions, it is possible to isolate quantities of substance suitable for obtaining an infrared spectrum. Additional
elution and purification procedures appropriate for IR work will
be discussed here.

Percival and Griffiths (59) prepared thin layer plates with
a silver chloride layer. Different organic compounds, most of
which were colored dyes, were separated on these plates, and their
infrared spectra were determined *in situ* using an infrared Fourier
transform spectrometer. This method is faster than other methods
but special layers and instruments must be used. This instrument
may also have reduced the detection limits somewhat.

Nash et al. (60) identified the pesticide rotenone in a TLC
chromatogram of technical grade rotenone using potassium bromide
(KBr) microdisks.

McCoy and Fiebig (61) eluted 50-100 µg amounts of separated
substance into the tip of a Pasteur pipet in a manner similar to
the one described in the previous section (Figure 14.4). The
method became cumbersome when the eluting solvent was to be replaced with carbon disulfide or carbon tetrachloride so the
spectrum could be run.

Hayden et al. (62) constructed a glass extraction apparatus
similar to a micro Soxhlet extractor for eluting small TLC zones.
The sorbent containing the sample is supported by a small pad of
glass wool rather than an extraction thimble.

A micro collection-elution device also fashioned from a
Pasteur pipet was developed by Clemett (63) and is shown in Figure 14.5. The body of the Pasteur pipet may be cut to any length
to accommodate the desired amount of sorbent material. The tip
end, plugged with glass wool, is connected to a carefully regu-
lated vacuum source, and the desired zone is sucked through the
capillary into the pipet. The plugged pipet may also be placed
into a side-armed test tube, the mouth of which contains a stopper and the capillary tube going into the pipet body. The vacuum
source is then connected to the side arm. Vacuum control is
easier this way, and the apparatus is not as fragile.

After collection, the pipet becomes a miniature chromatographic column through which the sample is eluted with a small
volume of pure solvent. Clean IR spectra have been obtained with
as little as 20 µg of sample, the eluate being dropped onto a
small amount of KBr powder that was pressed into a microdisk after
the solvent was evaporated. For mass spectrometry, the eluate

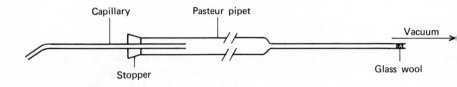

Figure 14.5.   Micro vacuum collection device for TLC zones.

was dropped slowly onto the sample probe tip and the solvent was
allowed to evaporate. Spectra were obtained from 5 µg of sample.
For gas chromatography, the elute could be collected in a micro
sample tube, the solvent evaporated under nitrogen, and the residue redissolved in an appropriate amount of suitable solvent for
injection. If evidence of contamination should result from this
method, it may then be desirable to fill the plugged tip of the
pipet with finely powdered KBr to prevent the passage of sorbent
material fines.

Garner and Packer (64) described the Wick-Stick method, kits
for which are available commercially from Harshaw Chemical Co.
The Wick-Stick is a triangle of pressed KBr 25 mm high, 8 mm wide
at the base, and 2 mm thick. For use, the triangle is inserted
into a support holder and placed in a small cylindrical glass
vial, all of which are supplied in the kit. The triangle must
stand erect and not touch the glass. Place the TLC layer sorbent
containing the sample to be eluted for the IR into the bottom of
the vial, using a funnel so as not to touch the triangle. If a
liquid sample is available, carefully run it down the side of the
vial. Add a small volume of solvent for elution and put on the
metal dust cap. The elution solvent dissolves the sample out of
the sorbent material and travels up the Wick-Stick by capillary
action. The sample material concentrates at the top of the triangle in the tip as the solvent evaporates out the hole in the dust
cap. Sorbent material is filtered out by the lower portion of
the triangle. Small samples may be effectively concentrated by
adding one or two more volumes of pure solvent to the vial to repeat the process. After elution, carefully remove the triangle
with forceps and place it on a clean inert surface such as a
piece of glassine weighing paper. Cut off the tip with a spatula
or scalpel and form it into a KBr pellet in the usual manner. Recovery is usually of the order of 50-80%, and spectra have been
obtained with 10 µg of separated substance.

Working with amino acid phenylthiohydantoins, Murray and

Smith (65) developed the following simple elution procedure for IR analysis. The separated area removed from the TLC plate is allowed to stand overnight with 1 ml of spectral grade 1,2-dichloroethane in a 100×14 mm tube with a Teflon-lined screw cap. The tube is centrifuged at about 300 rpm, and 80% of the supernatant is transferred to a clean microcentrifuge tube with a Pasteur pipet. Care is taken not to disturb the sorbent on the bottom of the tube. The transferred eluate is evaporated under nitrogen until a volume of 5-25 µl remains. This is spotted onto a silver chloride plate or a KBr disk with a microsyringe. The solvent is carefully allowed to evaporate. Spectra are recorded after horizontal and vertical adjustment of the sample holder in the beam condenser to attain maximum absorbance. The baseline is set by attenuating the reference beam, and water vapor bands are eliminated by adjustment of the balance control. All spectra are recorded with the slit program designated as "normal."

Spectral grade dichloroethane was used because spectral grade acetone and methanol contained a ketonic residue (0.0003-0.0005%) that yielded an intense spectrum upon evaporation of 1 ml volumes. The dichloroethane must not be exposed to air and light for extended periods, or it becomes unsuitable for this procedure, apparently due to oxidation and/or polymerization reactions.

Rice (66) developed a direct-transfer method that reduced handling of the area to be eluted to an absolute minimum. This was done by marking the area around the sample spot in the shape of a teardrop. An area around this teardrop was then scraped away and discarded for the next step. Powdered KBr (10-20 mg) was placed in this cleared area at the tip of the sample spot teardrop. Pure solvent for elution was added dropwise from a syringe onto the spot to elute the sample into the KBr. Amos (58) used this procedure but found "that considerable losses of solute occurred, particularly with colorless compounds. Furthermore, it was found to be suitable only for well resolved spots." The latter point is evident because of the area around the teardrop spot that must be cleared to allow for the KBr.

A variation of this technique was used by deKlein (67). The difference consisted in arranging the KBr powder in a semicircle in the clear area around the zone to be eluted. The powder is placed 1-2 mm away from the sample area and does not touch it as in the Rice method. The major disadvantage to this is that no portion of the KBr acts as a filter for the sorbent material, as could be the case with the Rice method. The whole of the KBr must be used to form the pellet for the IR. Szekely (68) has noted that a good recovery is possible with the semicircle arrangement of KBr, but the IR spectra obtained through such a procedure exhibited severe contamination, particularly when polar eluting solvents were used.

Amos (58) has developed an elution technique that he prefers. In its entirety it is as follows: 40 μl of a 5% solution of the sample is applied as a 2 cm band on a 5×20 cm silica gel HR plate 750 μ thick.   Two μl of a 1% solution of the same sample is chromatographed alongside and used as a reference.   The chromatogram is developed using only very pure solvents in the mobile phase, and the reference lane is sprayed with iodine solution (a suitable one is listed in Chapter 7) so that the location of the separated components from the large sample band can be ascertained.   The zone desired for elution is scraped off with a microspatula and placed on top of a small amount of powdered KBr that is tamped down the cone joint of an 18 gauge syringe needle, as in Figure 14.6.   The KBr serves to filter out the fines from the sorbent material.   A 1 ml syringe is filled with high purity acetone and connected to the needle, and the resulting solute is eluted drop by drop onto a 10 mg pile of powdered KBr.   Each drop is allowed to evaporate before addition of the next drop.   About 20 drops is usually sufficient to transfer the sample out of the sorbent into the KBr.   The eluted sample and the KBr are mixed with a microspatula and pressed into a 1.5 mm diameter disk.   Excellent spectra were obtained on 5-30 μg of solute, using a Perkin-Elmer 521 IR spectrophotometer.

Thin layer separation was combined with GC and IR for the identification of nitro compounds (69).

14.6 MISCELLANEOUS COMBINATIONS

Irvine and Anderson (70) defined the technique of autotransfer chromatography (ATC) as chromatography in two or more dimensions, where the stationary phase is changed and there is facile total transfer of sample components.   Such an example would be the transfer from TLC to paper chromatography.   This technique may provide additional information about the compounds being separated by virtue of the direct coupling of the two procedures.   Using this definition, the direct coupling of the effluent of a gas chromatograph to a TLC plate would be an autotransfer technique. The TLC/PC coupling provides a good deal of flexibility for the separation because the nature of the TLC layer and its mobile phase is variable, the solvent used for the actual transfer from the TLC layer to the paper is variable, and the mobile phase used for the development of the paper is still another variable that may be changed to fit the separation.

Irvine and co-workers (71) used this technique and combined it with elution from the paper and mass spectrometry for the identification and characterization of pyrroles and indoles from urine extracts.

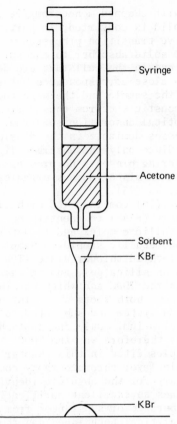

Figure 14.6   Syringe elution apparatus.

Differential thermal analysis (DTA) has been combined with
TLC for the identification of dicarboxylic acids at the 20-60 μg
level (72).

One of the newest techniques used in combination with TLC is
photoacoustic spectrometry. Rosencwaig and Hall (73) note that
the use of conventional spectrometric methods for the location of
compounds on TLC plates has not proved to be reliable, especially
in the UV wavelength region, because the sorbent material is usually too opaque to perform transmission spectrometry and too highly scattering to carry out reflection spectrometry. Photoacoustic
spectrometry is able to locate compounds *in situ* on the plate.

In this technique, a solid sample is placed inside a closed
cell containing air and a sensitive microphone. The sample is

then illuminated with chopped monochromatic light, and any light absorbed by the solid is converted, in part or in whole, into heat by nonradiative transition processes. The resultant periodic heat flow from the solid absorber to the surrounding gas creates pressure fluctuations in the cell that are detected by the microphone. There is a close correspondence between the amount of light absorbed by the sample and the magnitude of the acoustic signal. A photoacoustic spectrum thus corresponds, qualitatively at least, to an optical absorption spectrum, provided that the nonradiative processes dominate in the dissipation of the absorbed light energy. Since only the absorbed light is converted into sound, light scattering presents no problems. Rosencwaig has obtained optical data on miscellaneous organics, inorganic materials (74), and hemoproteins (75).

Issaq and Barr (76) combined TLC with flameless atomic absorption spectrometry (FAAS) to identify an inorganic compound in an impure organometallic complex and to determine the recovery and purity of the organometallic samples. Samples of the organometallic tellurium diethyldithiocarbamate (TDDC) always produced two spots upon TLC on silica gel, and it was impossible to determine which spot was the TDDC and which was the impurity. To distinguish between them, both spots were eluted from the plate using an automatic elution system and subjected to atomic absorption spectrometry for tellurium. Only one spot showed the presence of tellurium, and this therefore was the TDDC.

The many examples cited in this chapter show how the separation methods of thin layer chromatography complement the analytical methods necessary for the absolute identification of a substance. TLC provides an excellent purification method for separating a sought substance or substances from other contaminants in the sample. Analytical techniques can then be used to identify the separated substance.

REFERENCES

1.   J. Janak, J. Chromatogr., 48, 279 (1970).
2.   J. Janak, in Pharmaceutical Applications of Thin Layer and Paper Chromatography, K. Macek, Ed., Elsevier, Amsterdam, 1972.
3.   H. Schlitt and F. Geiss, J. Chromatogr., 67, 261 (1972).
4.   P. R. Boshoff, B. J. Hopkins, and V. Pretorius, J. Chromatogr., 126, 35 (1976).
5.   L. R. Snyder, J. Chromatogr. Sci., 7, 360 (1969).
6.   J. C. Touchstone and M. F. Dobbins, J. Steroid Biochem., 6, 1389 (1975).

7.   J. C. Touchstone, T. Murawec, O. Brual, and M. Breckwoldt, Steroids, *17*, 285 (1971).
8.   M. Breckwoldt, T. Murawec, and J. C. Touchstone, Steroids, *17*, 305 (1971).
9.   H. H. Wotiz and S. C. Chattoraj, Anal. Chem., *36*, 1466 (1964).
10. W. P. Collins and I. F. Sommerville, Nature, *208*, 836 (1964).
11. A. Pinelli and M. L. Formento, J. Chromatogr., *68*, 67 (1972).
12. A. Vermeulen, Clin. Chem. Acta, *34*, 223 (1971).
13. F. L. Berthou, L. G. Bardou, and H. H. Floch, J. Steroid Biochem., *3*, 819 (1972).
14. M. Luisi, C. Fassora, C. Levanti, and F. Franchi, J. Chromatogr., *58*, 213 (1971).
15. E. Watson and S. M. Kalman, J. Chromatogr., *56*, 209 (1971).
16. T. A. Smith, Anal. Biochem., *33*, 10 (1970).
17. J. C. Hoffsommer, J. Chromatogr., *51*, 243 (1970).
18. A. Copin, M. Severin, and J. Evrard, J. Chromatogr., *68*, 89 (1972).
19. C. G. Begg and M. R. Grimmett, J. Chromatogr., *73*, 238 (1972).
20. P. J. Bloom, J. Chromatogr., *75*, 261 (1973).
21. J. R. Evand, Clin. Chim. Acta, *42*, 343 (1972).
22. R. T. Nuttall and B. Bush, Analyst (London), *96*, 875 (1971).
23. P. W. Albro, L. Fishbein, and J. Fawkes, J. Chromatogr., *65*, 521 (1972).
24. R. M. Pfister, P. R. Dugan, and J. I. Frea, Science, *166*, 878 (1969).
25. J. L. Radomski and A. Rey, J. Chromatogr. Sci., *8*, 108 (1970).
26. H. V. Street, J. Chromatogr., *48*, 291 (1970.
27. S. J. Mule, J. Chromatogr., *55*, 255 (1971).
28. D. Sohn and J. Simon, Clin. Chem., *18*, 405 (1972).
29. E. E. Stoddard, J. Assoc. Offic. Anal. Chem., *55*, 1081 (1972).
30. O. S. Privett and E. C. Nickell, J. Am. Oil Chemists' Soc., *39*, 414 (1962).
31. O. S. Privett and E. C. Nickell, J. Lipid Res., *4*, 208 (1963).
32. O. S. Privett, M. L. Blank, and O. Pomanus, J. Lipid Res., *4*, 260 (1963).
33. O. S. Privett and E. C. Nickell, J. Am. Oil Chemists' Soc., *41*, 72 (1964).
34. R. Kaiser, in *Thin-Layer Chromatography, A Laboratory Handbook*, E. Stahl, Ed., Springer-Verlag, New York, 1969.
35. J. Janak, J. Chromatogr., *48*, 279 (1970).
36. J. Janak, in *Pharmaceutical Applications of Thin-Layer and Paper Chromatography*, K. Macek, Ed., Elsevier, Amsterdam, 1972.
37. J. Janak, Nature, *195*, 696 (1962).
38. J. Janak, J. Gas Chromatogr., *1*, 20 (1963).
39. J. Janak, J. Chromatogr., *15*, 15 (1964).

40. N. Ruseva-Atanasova and J. Janak, J. Chromatogr., *21*, 207 (1966).
41. B. Casu and L. Cavallotti, Anal. Chem., *34*, 1514 (1962).
42. J. H. Tumlinson, J. P. Minyard, P. A. Hedin, and A. C. Thompson, J. Chromatogr., *29*, 80 (1967).
43. J. P. Minyard, J. H. Tumlinson, A. C. Thompson, and P. A. Hedin, J. Chromatogr., *29*, 88 (1967).
44. H. C. Curtius and M. Muller, J. Chromatogr., *32*, 222 (1968).
45. A. M. Humphrey, J. Chromatogr., *53*, 375 (1970).
46. J. C. Touchstone, Ed., *Quantitative Thin Layer Chromatography*, Wiley, New York, 1973.
47. T. Cotgreave and A. Lynes, J. Chromatogr., *30* 117 (1967).
48. F. B. Padley, J. Chromatogr., *39*, 37 (1969).
49. J. J. Szakasits, P. V. Peurifoy, and L. A. Woods, Anal. Chem., *42*, 351 (1970).
50. B. P. Lisboa, J. Chromatogr., *48*, 364 (1970).
51. J. R. Majer, R. Perry, and M. J. Reade, J. Chromatogr., *48*, 328 (1970).
52. M. Fetizon, in *Thin Layer Chromatography*, G. B. Marini-Bettolo, Ed., Elsevier, Amsterdam, 1964.
53. M. A. Schwartz, P. Bommer, and F. M. Vane, Arch. Biochem. Biophys., *121*, 508 (1967).
54. M. A. Schwartz, F. M. Vane, and E. Postma, Biochem. Pharmacol., *17*, 965 (1968).
55. A. A. Boulton and J. R. Mayer, J. Chromatogr., *48*, 322 (1970).
56. G. Szekely, J. Chromatogr., *48*, 313 (1970).
57. M. J. Rix, B. R. Webster, and I. C. Wright, Chem. Ind. (London), 452 (1969).
58. R. Amos, J. Chromatogr., *48*, 343 (1970).
59. C. J. Percival and P. R. Griffiths, Anal. Chem., *47*, 154 (1975).
60. N. Nash, P. Allen, A. Bevenue, and H. Beckman, J. Chromatogr., *12*, 421 (1963).
61. R. N. McCoy and E. C. Fiebig, Anal. Chem., *37*, 593 (1965).
62. A. L. Hayden, W. L. Brannon, and N. R. Craig, J. Pharm. Sci., *57*, 858 (1968).
63. C. J. Clemett, Anal. Chem., *43*, 490 (1971).
64. H. R. Garner and H. Packer, Appl. Spectry, *22*, 122 (1967).
65. M. Murray and G. F. Smith, Anal. Chem., *40*, 440 (1968).
66. D. D. Rice, Anal. Chem., *39*, 1906 (1967).
67. W. J. deKlein, Anal. Chem., *41*, 667 (1969).
68. S. Szekely, in *Pharmaceutical Applications of Thin Layer and Paper Chromatography*, K. Macek, Ed., Elsevier, Amsterdam, 1972, p. 104.
69. A. Copin, M. Severin, and J. Evrard, J. Chromatogr. *68*, 89 (1972).

70. D. G. Irvine and M. E. Anderson, J. Chromatogr., *20*, 541 (1965).
71. D. G. Irvine, W. Bayne, and J. R. Majer, J. Chromatogr., *48*, 334 (1970).
72. R. V. Mangravite, Anal. Chem., *40*, 250 (1968).
73. A. Rosencwaig and S. S. Hall, Anal. Chem., *47*, 548 (1975).
74. A. Rosencwaig, Opt. Commun., *7*, 305 (1973).
75. A. Rosencwaig, Science, *181*, 657 (1973).
76. H. J. Issaq and E. W. Barr, Anal. Chem., *49*, 189 (1977).

Appendix of Suppliers

ANALABS, INC.
P. O. Box 501
New Haven, Conn. 06473

Supplies, plates, sorbent material, standards, reagents

ANALTECH, INC.
75 Blue Hen Drive
Newark, Delaware 19711

Plates.

APPLIED SCIENCE LABORATO-
RIES, INC.
P. O. Box 440
State College, Pa. 16801

Supplies, plates, sorbent material, standards, reagents

J. T. BAKER CHEMICAL CO.
222 Red School Lane
Phillipsburg, N.J. 08865

Plates, sorbent material, reagents, solvents.

BIO-RAD LABORATORIES
32nd and Griffin Avenue
Richmond, Calif. 94804

Sorbent material.

BRINKMANN INSTRUMENTS, INC.
Cantiague Road
Westbury, N.Y. 11590

Distributors for E. Merck (EM) and Macherey, Nagel (MN), Brinkmann, and Desaga brand products. Supplies, plates, sorbent material.

CAMAG, INC. 16229 West Ryerson Road P. O. Box 183 New Berlin, Wi. 53151	Supplies, plates, sorbent material, instruments.
EASTMAN KODAK CO. Organic Chemicals Div. Rochester, N.Y. 14650	Supplies, plates, standards, reagents, solvents.
EM LABORATORIES, INC. 500 Executive Boulevard Elmsford, N.Y. 10523	Distributors of E. Merck products, plates, sorbent material.
GELMAN INSTRUMENT CO. 600 South Wagner Road Ann Arbor, Michigan 48106	Supplies, plates, standards, reagents.
HARSHAW CHEMICAL CO. 1945 East 97th Street Cleveland, Ohio 44106	Wick Stick for infrared spectroscopy.
ICN PHARMACEUTICALS, INC. 26201 Miles Road Cleveland, Ohio 44128	Distributors for Woelm brand adsorbents.
KONTES GLASS CO. Spruce Street Vineland, N.J. 08360	Supplies, plates, instruments.
NEW ENGLAND NUCLEAR 540 Albany Street Boston, Mass. 02118	Plates.
PIERCE CHEMICAL CO. Box 117 Rockford, Ill. 61105	Reacti-Vials, derivatization reagents.
PHARMACIA FINE CHEMICALS, INC. 800 Centennial Avenue Piscataway, N.J. 08854	Distributors of Sephadex layer and column material, supplies.
QUANTUM INDUSTRIES 341 Kaplan Drive Fairfield, N.J. 07006	Supplies, plates, standards, reagents.

REGIS CHEMICAL CO.   PMD Instrumentation, reagents.
8210 Austin Avenue
Morton Grove, Ill. 60053

SCHLEICHER AND SCHUELL, INC.   Plates, sorbent material.
Keene, N.H. 03431

SCHOEFFEL INSTRUMENT CORP.   Plate scorer, densitometer.
24 Booker Street
Westwood, N.J. 07675

SHANDON SOUTHERN INSTRUMENTS   Supplies, radioscanning instru-
515 Broad Street   ments.
Sewickley, Pa. 15143

SINDCO CORP.   Accessories
P. O. Box 341567
Coral Gables, Fla. 33134

SUPELCO, INC.   Plates, supplies, standards, re-
Supelco Park   agents.
Bellefonte, Pa. 16823

A. H. THOMAS CO.   Supplies, plates, standards, re-
Vine Street at Third   agents.
Philadelphia, Pa. 19105

WHATMAN, INC.   Distributors for Reeve Angel
9 Bridewell Place   brand, Whatman brand chromato-
Clifton, N.J. 07014   graphic media.

Index

A